SpringerBriefs in Applied Sciences and Technology

More information about this series at http://www.springer.com/series/8884

Chiow San Wong · Rattachat Mongkolnavin

Elements of Plasma Technology

Chiow San Wong
Plasma Technology Research Centre,
Physics Department, Faculty of Science
University of Malaya
Kuala Lumpur
Malaysia

Rattachat Mongkolnavin
Department of Physics, Faculty of Science
Chulalongkorn University
Bangkok
Thailand

ISSN 2191-530X ISSN 2191-5318 (electronic)
SpringerBriefs in Applied Sciences and Technology
ISBN 978-981-10-0115-4 ISBN 978-981-10-0117-8 (eBook)
DOI 10.1007/978-981-10-0117-8

Library of Congress Control Number: 2015957048

Printed on acid-free paper

This Springer imprint is published by SpringerNature
The registered company is Springer Science+Business Media Singapore Pte Ltd.

Preface

Plasma technology is often referred to as one of the critical technologies of the twenty-first century. In the developed countries such as the USA and Japan, intensive scientific effort and large funding are being spent on research to develop plasma technologies that are useful for industry. In recent years, newly developed nations such as Korea, China (including Taiwan), and Singapore have also joined the race in the quest to be at the frontiers in this area of technology development. The effect of such efforts on the national economy of these nations is obvious.

This monograph covers some fundamental aspects of plasma technology which the authors believe are important to help researchers who intend to start work in plasma technology development. These include basic properties of plasma, methods of plasma generation, and basic plasma diagnostic techniques. The treatments are more experimental than theoretical, although some discussions on theoretical background will be given where appropriate. It is also not the intention of the authors to provide a complete and comprehensive treatment of all the topics covered. The more advanced readers may find the contents of this monograph too "elementary," which is precisely its intended aim. This monograph is for the beginners, not for the experts.

Also included in this monograph are discussions on several plasma devices with potential applications in industry including pulsed plasma radiation sources, low-temperature plasmas such as glow discharge and dielectric barrier discharge. The principles of the operation of these devices and some results of the studies carried out by the authors and their teams at the Plasma Technology Research Centre of University of Malaya and at the Physics Department of Chulalongkorn University will be reviewed and discussed.

Chiow San Wong
Rattachat Mongkolnavin

Acknowledgments

The authors are grateful to University of Malaya and Chulalongkorn University for providing research grants and facilities for them to carry out research in the area of plasma technology and also for providing job opportunity to them as well as supporting their collaboration. They also acknowledge the contribution and cooperation of group members, collaborators, and students in all the projects they have been involved. Particularly, the technical contribution of Mr. Jasbir Singh is most valuable toward the success of all the projects carried out by them. In addition, they would also like to thank the American Institute of Physics, Elsevier, Cambridge University Press and The Japan Society of Applied Physics for permission to use some figures from their publications as listed below:

Chin OH and Wong CS (1989) "A Simple Monochromatic Spark Discharge Light Source". *Rev. Sci. Instrum.* **60**: 3818–3819. © 1989, AIP Publishing LLC.

Wong CS, Woo HJ and Yap SL (2007) "A Low Energy Tunable Pulsed X-Ray Source Based On The Pseudospark Electron Beam". *Laser & Particle Beams* **25**: 497–502. © 2007, Cambridge University Press.

Chan LS, Tan D, Saboohi S, Yap SL, Wong CS (2014) "Operation Of An Electron Beam Initiated Metallic Plasma Capillary Discharge". *Vacuum* **103**: 38–42. © 2014, Elsevier.

Wong CS, Choi P, Leong WS and Jasbir S (2002) "Generation of High Energy Ion Beams from a Plasma Focus Modified for Low Pressure Operation". *Jpn. J. Appl. Phys.* **41**: 3943–3946. © 2002, JSAP.

Contents

Chapter 1
Basic Concepts in Plasma Technology

Abstract Some of the fundamental basic concepts useful for the understanding of plasma will be introduced and explained in this chapter. These include particle collision and the fundamental processes that may occur as a consequence of collision between particles, the concept of Debye shielding, plasma sheath formation at the surface of object placed inside plasma, particle oscillation. The criteria of plasma and the particle transport due to electric and magnetic fields as well as density gradient will be discussed briefly.

Keywords Plasma · Plasma properties

1.1 Plasma—The Fourth State of Matter

In the simplest term, we may say plasma is the fourth state of matter. At room temperature, there are matters that exist in the solid state, some in the liquid state and some in the gaseous state. Each kind of matter can also exist in all the three states if it is heated to temperature above room temperature and cooled to below room temperature. For example, water is in the liquid state at room temperature and atmospheric pressure. If water is cooled to below 0 °C, much lower than room temperature, it will turn into ice—the solid state of water. On the other hand, when water is heated to 100 °C, it turns into its gaseous state—steam (Fig. 1.1).

Now consider what will happen if the steam is heated to temperature much higher than 100 °C. Some of the water molecules (H_2O) may be dissociated into hydrogen and oxygen atoms. If the temperature is high enough, the hydrogen and oxygen atoms may even be ionized to form positive ions and electrons. This "ionized" state of water is the fourth state of matter, or it is called the plasma state if it satisfies certain criteria. We will discuss about these criteria in more details later.

Inside the plasma, there are electrons, ions of various charge states, neutral atoms and molecules. These particles move around inside the plasma with kinetic energy. The particles may exchange energy when they collide with each other. The collision can be either elastic or inelastic.

C.S. Wong and R. Mongkolnavin, *Elements of Plasma Technology*,
SpringerBriefs in Applied Sciences and Technology,
DOI 10.1007/978-981-10-0117-8_1

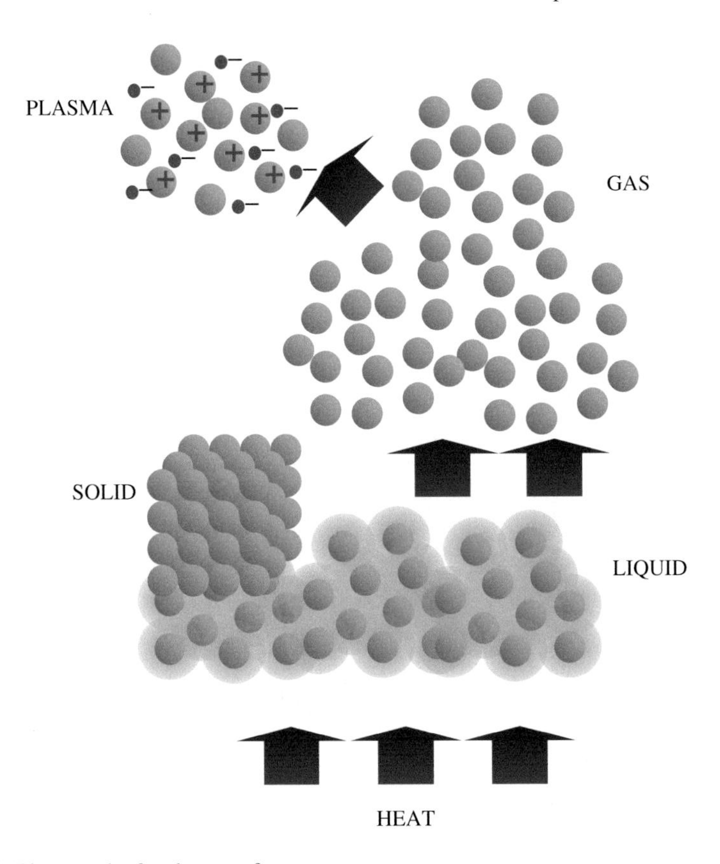

Fig. 1.1 Plasma—the fourth state of matter

During an elastic collision, the particles only exchange kinetic energy. The magnitudes and directions of the velocities of both colliding particles may be changed after the collision. However, during an inelastic collision, the internal energy of the colliding particles may be changed. This leads to the occurrence of various types of processes such as excitation and ionization. In particular, ionization gives rise to the production of new charged particles and hence causing the number of charged particles to increase.

1.2 Collision

1.2.1 Elastic Collision

Consider the case of collision between 2 spherical particles with mass m_1 and m_2 respectively (Fig. 1.2).

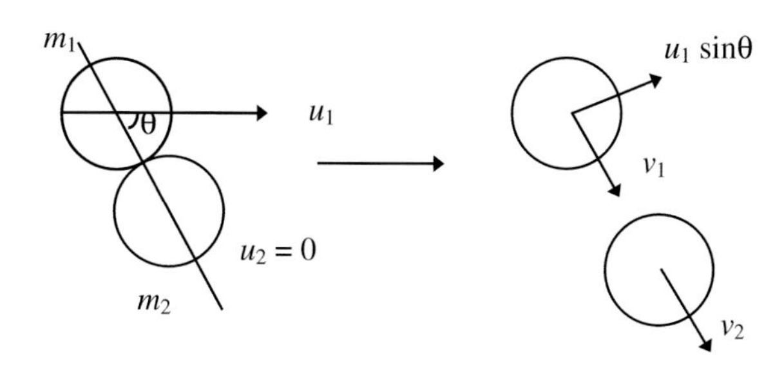

Fig. 1.2 Collision between particles

Assume that before collision, m_2 is at rest while m_1 is moving with velocity u_1. When they collide, at the point of contact, u_1 makes an angle of θ with the line joining the centers of the two colliding particles. After the collision, the velocity of m_1 becomes $\sqrt{v_1^2 + u_1^2 \sin^2 \theta}$, while m_2 will now start moving at a velocity of v_2 in the direction along the line joining the centers of the two colliding particles.

Considering conservation of momentum before and after the collision,

$$m_1 u_1 \cos\theta = m_1 v_1 + m_2 v_2$$
$$m_1 u_1 \sin\theta - \text{ unchanged}$$
$$v_1 = u_1 \cos\theta - \frac{m_2}{m_1} v_2$$

Similarly, considering conservation of energy before and after the collision,

$$\frac{1}{2} m_1 u_1^2 = \frac{1}{2} m_1 \left(v_1^2 + u_1^2 \sin^2 \theta \right) + \frac{1}{2} m_2 v_2^2$$
$$\left(\frac{v_2}{u_1}\right)^2 = \left(\frac{2 m_1 \cos\theta}{m_1 + m_2}\right)^2$$

The fraction of energy transferred during the collision:
δ = kinetic energy of m_2 after collision/initial kinetic energy of m_1

$$= \frac{\frac{1}{2} m_2 v_2^2}{\frac{1}{2} m_1 u_1^2} = \frac{m_2}{m_1} \frac{v_2^2}{u_1^2} = \frac{m_2}{m_1} \frac{4 m_1^2 \cos^2 \theta}{(m_1 + m_2)^2} = \frac{4 m_1 m_2}{(m_1 + m_2)^2} \cos^2 \theta.$$

Consider

$$\text{(i)} \quad m_1 = m_2 \Rightarrow \delta = \cos^2\theta, \delta_{\max} = 1$$
$$\text{(ii)} \quad m_1 \ll m_2, \quad \therefore m_1 + m_2 \approx m_2.$$
$$\Rightarrow \delta \approx 4 \frac{m_1}{m_2} \cos^2 \theta, \quad \text{hence } \delta_{\max} = \frac{4 m_1}{m_2} \ll 1.$$

Since $m_1 \ll m_2$, the value of δ is always small. This is the case for electron-atom or electron-ion collision. We see that with elastic collision, the tracks of the electrons will be deflected when they collide with atoms or ions, with only a small fraction of their energy is transferred to the atoms or ions. In other word, they will be scattered. This is what happen for a gas at low or not so high temperature. When the temperature is increased, the particles are moving with higher kinetic energy and they will collide harder. When the collision is sufficiently hard, energy will be transferred into the internal energy of the target particle and the collision is said to be inelastic.

1.2.2 Inelastic Collision

For inelastic collision, the momentum is still conserved as in the case of elastic collision, where

$$m_1 u_1 \cos\theta = m_1 v_1 + m_2 v_2$$
$$m_1 u_1 \sin\theta - \text{unchanged.}$$

But the energy equation has to be written as

$$\frac{1}{2} m_1 u_1^2 = \frac{1}{2} m_1 (v_1^2 + u_1^2 \sin^2\theta) + \frac{1}{2} m_2 v_2^2 + \Delta U$$

where ΔU is the energy transferred into the internal energy of m_2 during the collision.

ΔU can be shown to be: $\Delta U = m_2 u_1 v_2 \cos\theta - \frac{m_2 (m_1 + m_2) v_2^2}{2 m_1}$.

Differentiate this with respect to v_2 gives

$$\frac{d(\Delta U)}{dv_2} = m_2 u_1 \cos\theta - \frac{m_2 (m_1 + m_2)}{2 m_1} (2 v_2)$$

For ΔU = maximum, let $d(\Delta U)/dv_2 = 0$

$$\Rightarrow v_2 = \left(\frac{m_1}{m_1 + m_2}\right) u_1 \cos\theta$$
$$\therefore (\Delta U)_{\max} = \frac{1}{2} \left(\frac{m_1 m_2}{m_1 + m_2}\right) u_1^2 \cos^2\theta$$

The maximum energy transferred to the internal energy of the target particle is given by:

$$\delta = \left(\frac{m_2}{m_1 + m_2}\right) \cos^2\theta.$$

For $m_1 \ll m_2, \delta = \cos^2\theta,\ \delta_{\max} = 1.$

This means that during an inelastic collision, all the energy of the electron (colliding particle) can be transferred to the internal energy of m_2, which may be ion or atom.

With the increase in the internal energy, the atom or ion may undergo changes in its electronic configuration leading to processes such as excitation and ionization.

1.3 Collision Cross-Section

In the consideration of collision between electron and atom or ion an important question to ask is: What is the probability that a collision will occur? And subsequently, what is the probability that the collision will lead to a particular process? This is expressed in terms of the cross-section. For collision between electron and atom, the cross-section can be simply defined by the size of the atom. Hence the cross-section for electron-atom collision is given by $\sigma_{e-a} = \pi a_o^2$, where a_o is the radius of the atom. For collision between charged particles, because of the coulomb effect, the particles may be able to exchange energy without physical contact. we may define an impact parameter b which is the distance of closest approach between the colliding particles. In the particular case when the incoming particle is deflected at 90° with respect to its original path, the coulombic potential energy between them (attractive for e-i or repulsive for e-e and i-i collisions) at the point of closest approach can be shown to be equal to twice the kinetic energy of the incoming particle at infinity. That is,

$$\frac{Z_1 Z_2 e^2}{4\pi\varepsilon_o b_o} = 2 \times \frac{1}{2} m u_1^2$$

where b_o is the distance of closest approach for 90° deflection. This gives

$$b_o = \frac{Z_1 Z_2 e^2}{4\pi\varepsilon_o (3kT)},$$

where we have taken $\frac{1}{2} m u_1^2 = \frac{3}{2} kT$. For collision between charged particles, the cross-section is written as $\sigma = \pi b_o^2$. Hence

$$\sigma = \pi \left[\frac{Z_1 Z_2 e^2}{4\pi\varepsilon_o (3kT)} \right]^2.$$

This expression, however, applied only to a slightly ionized plasma. For fully or strongly ionized plasma,

$$\sigma = 5.7\ell n\Lambda\pi\left[\frac{Z_1Z_2e^2}{4\pi\varepsilon_o(3kT)}\right]^2,$$

where $\ell n\Lambda \approx 10$ for fully or strongly ionized plasma.

With reference to the cross-section, we can define the mean free path of collision which is the average time interval between 2 consecutive collisions as $\lambda = \frac{1}{n_2\sigma}$ and the mean free time as $\tau = \frac{1}{n_2 u_1 \sigma}$ where n_2 is the number density of the target particles and u_I is the velocity of the incoming particle.

1.4 Fundamental Processes in a Plasma

As a consequence of collisions between electron and the atom or ion inside the plasma, various processes may occur. Four of the most elementary processes are:

i. Scattering $e + A \rightarrow A + e$
This is caused by elastic collision where the colliding electron will transferred a small fraction of its kinetic energy to the atom or ion. However, its direction of motion will be changed.

ii. Excitation $e + A \rightarrow A^* + e$
This occurs when electron of sufficient energy collides at an atom or ion inelastically. Part of its kinetic energy is absorbed by an inner shell electron of the atom or ion so that this inner shell electron is raised to a higher energy level and the atom or ion thus becomes excited.
Most of the excited states have short life time and they will decay back to their original levels by emitting a photon equivalent to the energy difference. This process is called de-excitation or relaxation by spontaneous emission.

iii. Ionization $e + A \rightarrow A^+ + 2e$
With sufficiently high energy, the colliding electron may transfer enough energy into the internal energy of the atom or ion to release one of the bound electrons. The atom or ion thus becomes one charge state higher and it is said to be ionized.
It is through the ionization process that new charged particles (electrons) are produced in the plasma.

iv. Recombination $e + A \rightarrow A + (hv)$
In this case the electron colliding with the ion may be captured and it occupies the vacancy inside the ion to change its charge state to one level lower than previously. A photon may be emitted when the electron releases its excess energy.
The question of which process is probable can be answered by referring to the cross section of the process which is expressed as probability of its occurrence at various energy of the colliding electron. An example of the cross sections for processes (i), (ii) and (iii) for argon gas is shown in Fig. 1.3.

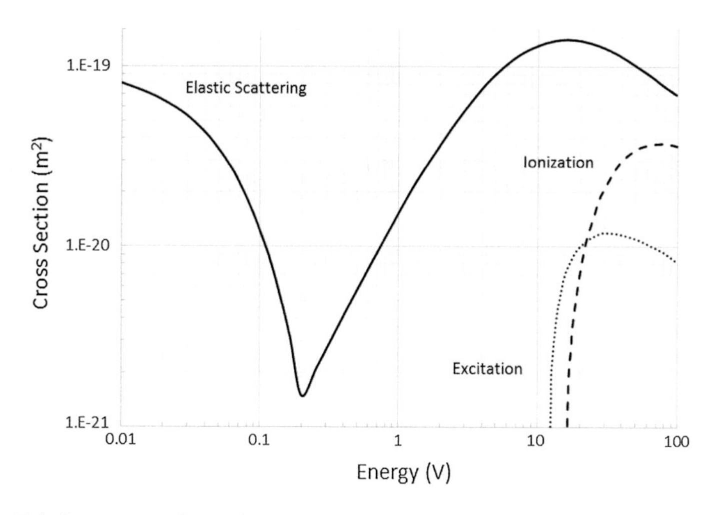

Fig. 1.3 Relative cross-sections of scattering, excitation and ionization processes

1.5 Some Consideration of Thermodynamic Properties of Plasma

Let us consider a gas at sufficiently low temperature when it can be considered to consist of predominantly neutral particles. The types of species present may consist of atoms (and/or molecules), ions and electrons. Under such a condition, assuming thermodynamic equilibrium, the ideal gas law may be applied. If we define α_i to be the fractional population of an ionic specie so that

$$\alpha_i = \frac{N_i}{N_t} \quad \text{where } N_t = \sum_{j=0}^{Z} N_j.$$

Hence

$$\alpha_o = \frac{N_o}{N_t}$$

is the fractional population of neutral particles, and

$$\alpha_1 = \frac{N_1}{N_t}$$

is the fractional population of singly ionized species and so on. For the gas at low temperature (say near to the room temperature), we expect

$$\alpha_0 \approx 1,\ \alpha_{1,}\alpha_{2,} \approx 0$$

From kinetic theory, the pressure p (specific) of the gas is related to the temperature T and particle number density n as

$$p = nkT = \frac{N}{V}kT = \rho RT,$$

where V is the volume of the gas, ρ is the mass density, and R is the gas constant. This is the ideal gas law. For an ideal gas, the specific heat ratio γ is given by

$$\gamma = \frac{f+2}{f} = \frac{5}{3},$$

since an ideal gas is assumed to have three degree of freedom ($f = 3$). When the gas is heated up (for example by electrical discharge), the kinetic energy of the particles inside the gas increases and their collisions will also be increasingly "hard" and hence inelastic. This is particularly effective when the colliding particles are electrons, as we have discussed earlier, although their fractional population is relatively small. This gives rise to transfer of energy into the internal energy of the target particles which may be atoms or ions. There is thus an increase in the degree of freedom which means f is getting larger. γ will then drop and approaching unity.

With the occurrence of inelastic collision, the cross section for ionization also increases and new charged particles (ions and electrons) will be produced. At a particular temperature, only a certain number (not more than 5 or 6) of species will be dominant. The distribution of fractional populations of the various species in a gas heated to elevated temperature can be described by either the Local Thermodynamic Equilibrium (LTE) Model or the Coronal Equilibrium (CE) Model. The choice of which model is appropriate depends on the density of the gas. In general, LTE model is more appropriate for a high density plasma while the CE model is more suitable for a low density plasma. It is often difficult to decide on what density is high enough to be considered "high". We will come back to this topic again when we consider the radiation emission spectrum of the plasma.

The fact that when a gas is heated to higher temperature, inelastic collisions will lead to processes such as excitation and ionization (hence increasing degree of freedom) implies that the gas begins to deviate from ideal gas condition. Hence the conversion of a gas from the neutral state to the ionized state (plasma) is converting it into a real gas. For plasma taken as a real gas, the appropriate form of the equation of state is

$$p = \rho RTz,$$

where z is the Departure Coefficient (departure from ideal gas), and R is the gas constant.

For a plasma with ionic species α_0, α_1, $\alpha_2 \ldots \alpha_i$,

$$z = 1 + \sum_{j=1}^{i} j\alpha_j \quad \text{for atomic gas,}$$

$$z = 1 + \chi + \sum_{j=1}^{i} [(2j+1)\alpha_j] \quad \text{for diatom molecular gas,}$$

where χ is the fractional population of molecules.

When the plasma is heated to extremely high temperature until it is fully ionized (that means all electrons are stripped off from all the atoms), processes such as excitation and ionization will not occur anymore and the number of degree of freedom reduced back to 3 only. In this case the plasma consists of only 2 types of species, fully stripped ions and electrons and it can be considered to be ideal gas again.

1.6 Concept of Plasma Potential

One of the important characteristics of plasma is that charge neutrality is assumed to be maintained. This means that there is a balance of positive and negative charges. If we assume that the charge density is zero everywhere inside the plasma, then there should be no electrostatic field. However, the potential is not zero inside the plasma, although it is uniform.

The fact that the plasma potential is not zero is evident from two consequences:

1. When a stray point charge is dropped into the plasma, the charged particles inside the plasma will re-distribute themselves so as to shield the effect of the stray charge to wihin a small sphere (the so called Debye sphere) with the stray charge as the center having radius of a Debye length given by

$$\lambda_D = 6.9\sqrt{\frac{T_e}{n_e}}\,\text{cm}\ [T_e\text{: }^\circ\text{K}\ ;\ n_e\text{: cm}^{-3}]$$

This effect is called the Debye Shielding. This is determined by the electrons inside the plasma because it is usually the electrons that will re-distribute to produce the screening effect. When the stray charge is positive, the electrons will move towards the charge so that the population of electrons inside the Debye sphere is higher than that of the ambient. On the other hand, if the stray charge is negative, the population of electrons inside the Debye sphere will be lower than the ambient.

2. Imagine now that the stray charge is suddenly removed from the plasma. The particles will try to re-distribute back to the original situation and as a consequence they may be set into simple harmonic oscillation. Taking analogy of a mass m hanging on a spring. If the mass is pulled downward to make a displacement and then let go, the mass will tend to go back to its equilibrium position but it will not stop at the equilibrium position unless the system is critically damped. It will be set into oscillation (simple harmonic motion) instead. To better visualize the effect in a plasma we can refer to a static situation where all the charged particles are aligned in such a way a achieve charge neutrality. Taking the simplest case of 1-D, we have an infinite array of alternate positive and negative charges of equal magnitude. If an electron is displaced and then released, it will start to oscillate about its equilibrium position in simple harmonic motion with frequency given by

$$f_e \approx 9 \times 10^3 \sqrt{n_e}\,\mathrm{Hz}\ \left[n_e\text{: cm}^{-3}\right]$$

Note that f_e is only dependent on the density of electrons.

1.7 Criteria of Plasma

In a very broad term, we say a plasma is an ionized gas. Actually an ionized gas has to fulfil several criteria in order to be qualified as a plasma.

Criterion 1: $\lambda_D \ll L$

Here L is the "characteristic" dimension of the plasma. This criterion requires that if the plasma is perturbed by a stray charge, the effect of the stray charge should be shielded to within a distance of one Debye length which is much smaller than the characteristic dimension of the plasma.

Criterion 2: $N_D = \frac{4}{3}\pi\lambda_D^3 N \gg 1$

The number of particles inside the Debye sphere N_D must be sufficiently large, at least much greater than unity, say 100. This requires that the particle number density of the plasma is sufficiently high.

Criterion 3: $\omega_p \tau > 1$

$\omega_p = 2\pi f_e$ is the angular frequency of electron plasma oscillation while τ is the electron-atom or electron-ion collision mean-free-time. In other words, the plasma oscillation frequency must be higher than the collision frequency. This implies that between 2 collisions, the electron must have performed many oscillations. This is to ensure that the particle is able to reach equilibrium after each collision.

1.8 Effect of Boundary in Plasma

So far we have assumed a borderless and uniform plasma. What happen at the boundary of the plasma such as the chamber wall containing the plasma; or at the surface of an extended object (not a point charge) placed inside the plasma?

Let's say we start from an equilibrium situation where the plasma is uniform and homogenous. When an object (floating and not connected to anyway) is placed inside the plasma, the plasma will react and since the electrons are much lighter than the ions, they will reach the surface of the object first. This will cause the surface of the object to be "charged" up to a negative potential. However, as more electrons reach at the surface, the negative potential will also increase and tends to push away any subsequent electrons coming towards the surface. An equilibrium will soon be reached when the potential distribution in front of the surface becomes distorted from the plasma potential V_p as shown in Fig. 1.4. The potential at the surface of the object V_f is negative with respect to the plasma potential and it is often referred to as the floating potential of the object.

As a first approximation, it can be assumed that the potential will return to the plasma potential at a distance of λ_D (the Debye length) from the surface. In other words, a plasma sheath of thickness λ_D will be formed at the surface of the object placed inside the plasma.

The above simplified picture of plasma sheath formation is obtained from a static charge distribution model. Since the particles are moving around inside the plasma randomly, the ions will only be accelerated in the direction towards the surface when they get sufficiently close to the surface. Beyond that point, their velocities will gradually increase until

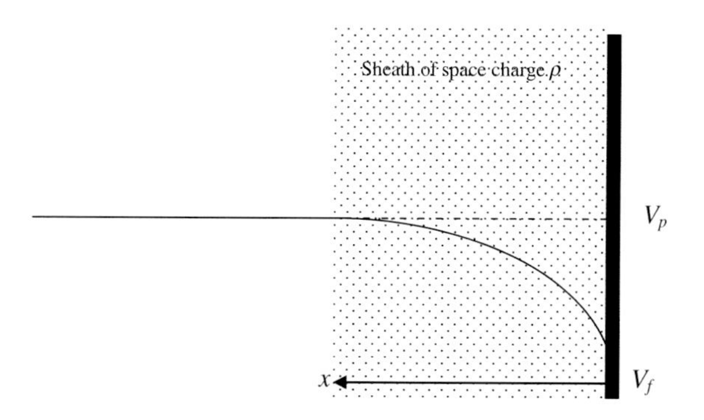

Fig. 1.4 Potential distribution at boundary in plasma—first approximation

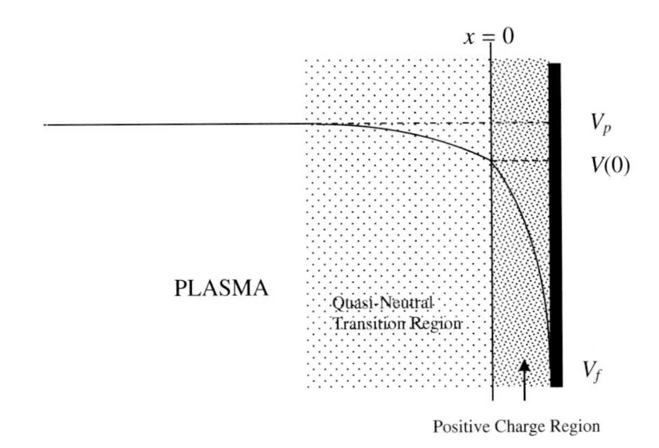

Fig. 1.5 Potential distribution at boundary in plasma—the Bohm model

$$u_o > \sqrt{\frac{kT_e}{m_i}}.$$

This is the Bohm criterion of the plasma sheath. A more accurate picture of the plasma sheath structure is shown in Fig. 1.5.

The ions are moving randomly in the unperturbed plasma. When they reach the quasi-neutral transition region they start to be accelerated towards the surface of the object. Beyond the point $x = 0$ in Fig. 1.5, all ions have achieved a velocity exceeding $\sqrt{\frac{kT_e}{m_i}}$. This is considered to be the boundary of the plasma sheath formed on the surface of the object.

1.9 Particle Transport Inside Plasma

So far we have considered charged particles in the plasma to be moving randomly due to their thermal velocity. How will the particles move under the influence of externally applied forces?

We will consider here two types of external forces that may influence the motion of the particles inside a plasma.

i. Motion of charged particles in an electric field
   For charged particles (specifically electrons) moving in vacuum under the influence of an electric field, we can write the equation of motion as (this is sometime referred to as Lorentz Model):

$$m_e \frac{dv_e}{dt} = eE \quad \text{which gives } v_e = \left(\frac{et}{m_e}\right)E.$$

Inside the plasma, due to collision, the motion of electrons will be affected. Instead of moving straight in the direction of the electric field as in the case of vacuum, the electron may change direction each time it collides with another particle. However, the overall direction of the electron is still expected to follow the direction of the electric field. The electrons are said to be drifting and the electron drift velocity will be proportional to the electric field. The electron drift velocity can be obtained by considering its motion between two consecutive collisions to be same as that of the vacuum case and taking the average effect over many collisions. Hence the electron drift velocity can be expressed as

$$v_e = \left(\frac{e\tau_{e-a}}{m_e}\right)E,$$

where $\tau_{e\text{-}a}$ is the electron-atom collision mean free time, the average time in between two collisions. The quantity $\mu_e = \left(\frac{e\tau_{e-a}}{m_e}\right)$ is called the electron mobility. This is in fact a measure of how the motion of the electron will be affected by collision. If the frequency of collisions is high, hence $\tau_{e\text{-}a}$ is short, then the electrons will have low mobility.

If we write the electron current density to be $J_e = n_e e v_e = n_e e \mu_e E$ and compared to Ohm's Law for electrical conduction $J_e = \sigma E$, then we see that we can define an equivalent electrical conductivity of the plasma as

$$\sigma = n_e e \mu_e = \frac{n_e e^2 \tau_{e-a}}{m_e},$$

Again, more frequent collision means lower conductivity.

Similarly consideration can be given to the ions. The direction of the drift will be opposite to that of electrons. However, due to the massiveness of the ions as compared to electrons ($m_p/m_e \sim 10^3$), $v_i \ll v_e$ under the influence of the same electric field. Hence in the consideration of particle transport in an electric field, it is common to assume the ions to be stationary.

ii. Motion of particles due to density gradient

The presence of density gradient inside the plasma may also change the motion of the particles from random to drifting in a particular direction. This phenomenon is called diffusion. The drift velocity created is directly proportional to the density gradient ∇n but is inversely proportional to the density n itself. Hence we write

$$v_d = -D\frac{\nabla n}{n}.$$

The minus sign is to show that the direction of the diffusion is opposite to the density gradient, that is from high density region to low density region.

Under the effect of density gradient, both electrons and ions will move in the same direction. However, due to the massiveness of the ions, they will lag behind and this results in charge separation. An induced electric field will be produced. This field is in such a direction to accelerate the ions but retard the electrons. Imagine that both electrons and ions are originally at the same point and start to diffuse outwards (spherical geometry) together. First the electrons will move much faster than the ions but soon charge separation occurs and the induced electric field will act to slow down the electrons but to accelerate the ions. An equilibrium will soon be established when the drift velocity of electrons and ions become the same. This situation is termed *ambipolar diffusion*.

Chapter 2
Methods of Plasma Generation

Abstract In this chapter, the generation of plasma by gaseous electrical discharge will be discussed. Townsend Theory of electrical breakdown of gases will be explained. Various types of discharge, including to corona discharge, glow discharge and arc discharge and the characteristics of the plasmas produced will be introduced. The electrical power sources used for the generation of these plasma including DC, AC, RF, microwave and pulsed capacitor discharge are introduced.

Keywords Electrical discharge · Pulsed discharge

2.1 DC Electrical Discharge

2.1.1 Electrical Breakdown

In its neutral form, a gas is an insulator and will not conduct electrical current when an electric field is applied across it, no matter how high is that electric field. However, some amount of stray charges are always present in the neutral gas. One such source of stray charges is the ionization of the gas particles by cosmic ray or any other background radiation from the environment. If the electric field is applied to the gas using electrodes, another source of stray charges (electrons) may come from the photoelectric effect at the cathode surface due to the absorption of UV photons. The presence of stray electrons is crucial in electrical discharge since they can be accelerated to high energy to produce ionizing collision and hence new charged particles. Without the presence of stray electrons, electrical discharge may not occur.

The simplest configuration to produce an electrical discharge through a gas is to apply a potential difference across a pair of parallel electrodes placed inside a chamber filled with the gas at a suitable pressure as shown in Fig. 2.1.

Consider an electron originated from the cathode due to the absorption of UV photons. With the presence of electric field, the electron will be accelerated to high enough energy for excitation or ionization when it collides with an atom. When an ionizing collision occurs, the colliding electron together with the new electron will be further accelerated by the electric field and they may produce further ionizing

C.S. Wong and R. Mongkolnavin, *Elements of Plasma Technology*,
SpringerBriefs in Applied Sciences and Technology,
DOI 10.1007/978-981-10-0117-8_2

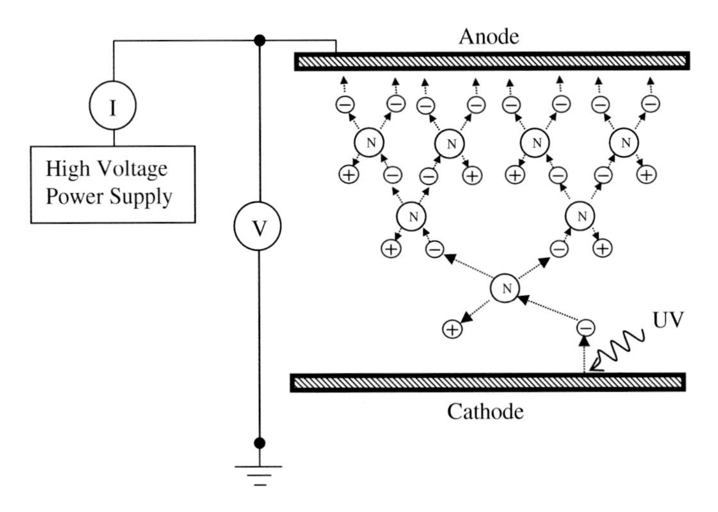

Fig. 2.1 Multiplication of charged particles leading to electrical discharge

collisions. Hence we see that the number of electrons will be multiplied as we progress from the cathode to the anode. This multiplication of charges (electrons) is described by the Townsend theory of gas discharge. The number of ionizing collisions per unit distance is called the First Townsend Coefficient, α. The relation of α with the gas pressure p and the applied electric field E have been established empirically for various gases. This is expressed as [1]:

$$\frac{\alpha}{p} = C\exp\left(-D\sqrt{\frac{p}{E}}\right) \tag{2.1}$$

for atomic gases, and

$$\frac{\alpha}{p} = A\exp\left(-B\frac{p}{E}\right) \tag{2.2}$$

for molecular gases.

The constants A, B, C and D are constants of the gas used. For example, for argon gas C = 29.22 cm^{-1} torr^{-1}, D = 26.64 (V cm^{-1} torr^{-1})$^{1/2}$; while for air A = 14.6 cm^{-1} torr^{-1}, B = 365 V cm^{-1} torr^{-1}.

Considering the process of ionizing collision by a flux of electrons originated from the cathode produced by UV absorption, and assuming that α is constant at constant E and p, Townsend's theory predicts that the electron flux will grow according to the simple relation of

$$F(x) = F_o\exp(\alpha x),$$

where F_o is the electron flux at the cathode surface expressed in terms of number of electrons per unit area.

For an anode-cathode distance of d, the electron flux finally collected by the anode is

$$F_a = F_o \exp(\alpha d).$$

The anode current density measured is $J_a = eF_a = J_o \exp(\alpha d)$. This can also be written in terms of the total current I_a if we assume that the cross-section of the discharge current remains the same. That is

$$I_a = I_o e^{\alpha d} \tag{2.3}$$

This is reasonably well obeyed by experimental data, as illustrated in the plot of ln I_a versus d as shown in Fig. 2.2.

The straight line does not pass through the origin, but intercept the y-axis at (ln $I_a - \alpha d_o$). This is due to the fact that the electrons must be accelerated for a distance of at least d_o before they acquire sufficient energy to produce ionizing collision. In that case the expression for the anode current density should be corrected as $I_a = I_o \exp[\alpha(d - d_o)]$. From Fig. 2.2, it can be seen that from the measurements of I_a at various electrode separation d, the value of α for the gas can be determined from the gradient of the straight line obtained. Note that for this experiment, when d is increased, the potential applied across the electrodes V must also be increased in order to keep $E = V/d$ constant. Hence the value of α obtained is for a particular E.

Figure 2.3 shows an example of the values of α/p as a function of E/p obtained experimentally for argon.

The plot can be converted to α/p as a function of $(p/E \times 1000)^{1/2}$ as shown in Fig. 2.4. The values of the constants C and D can then be determined from this plot.

In the above consideration, the electrons produced by ionization in the gas are assumed to be accelerated to the anode, leaving behind the heavy ions. For ions near to the cathode surface, they may collide at the surface and for sufficiently high electric field, the collision at the cathode surface by the ions may be energetic

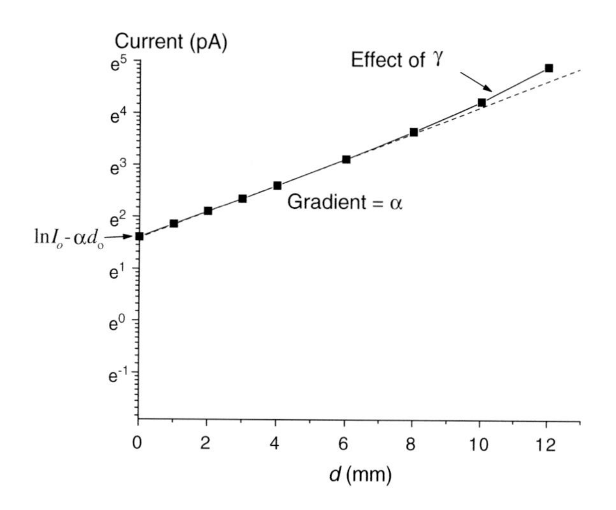

Fig. 2.2 Plot of lnI_e against inter-electrode distance d

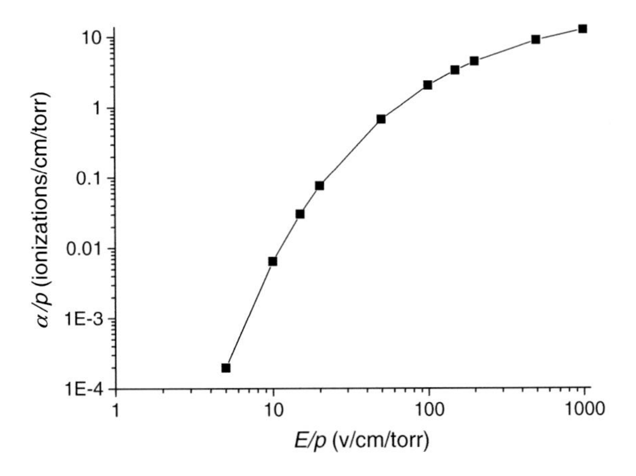

Fig. 2.3 Relationship between α/p and E/p

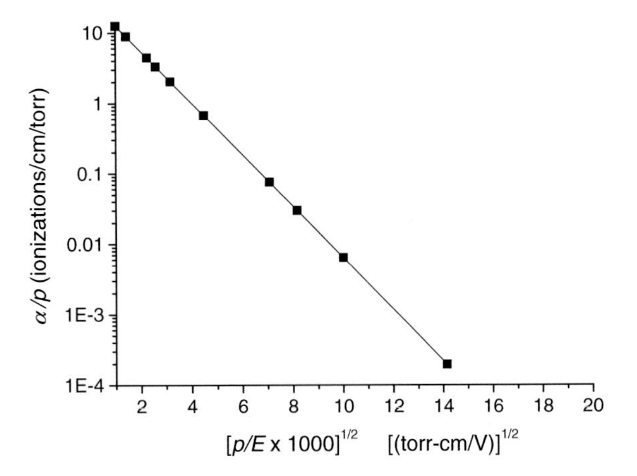

Fig. 2.4 Plot of α/p against p/E

enough to overcome the work function of the cathode material and releases electrons from it. The number of electrons released from the surface from each ion bombardment is called Townsend's Second Coefficient γ. This acts as an additional source of electrons which should be considered in Townsend's theory of gas discharge. The anode current is now given in the form

$$I_a = \frac{I_o \exp(\alpha d)}{1 - \gamma(e^{\alpha d} - 1)}. \tag{2.4}$$

Notice that for small d, $\exp(\alpha d) \sim 1$ so that the denominator will be ~ 1. This is observed experimentally. For small d, the plot of $\ell n\, I_a$ against d is a straight line. As d is increased to sufficiently large value, the straight line will begin to curve upwards, deviating from the straight line.

The release of electrons from the cathode surface is crucial in gas discharge. It is responsible for making the discharge "self-sustained" and do not rely on external source such as UV light.

From Eq. (2.4), if the denominator tends to zero, that is

$$1 - \gamma\left(e^{\alpha d} - 1\right) \rightarrow 0$$

then $I_a \rightarrow \infty$. This is the condition that an electrical breakdown has occured. A self-sustained discharge is obtained. The criterion that

$$\begin{aligned} &1 - \gamma\left(e^{\alpha d} - 1\right) = 0 \quad \text{or} \quad \gamma\left(e^{\alpha d} - 1\right) = 1 \quad \text{or} \quad \gamma\left(e^{\alpha d}\right) = \gamma + 1 \\ &\text{or} \quad \alpha d = \ell\text{n}\left(1 + \frac{1}{\gamma}\right) \end{aligned} \tag{2.5}$$

is referred to as the ***breakdown criterion*** of electrical discharge.

From the breakdown criterion, substitute for α/p from Eq. (2.1) for atomic gas, and writing $E = \frac{V_B}{d}$ at electrical breakdown, it can be shown that the breakdown voltage can be expressed as

$$V_B = \frac{D^2 pd}{\left[\ell\text{n}\left(\frac{Cpd}{\ell\text{n}\left(1 + \frac{1}{\gamma}\right)}\right)\right]} \tag{2.6}$$

for atomic gas, and similarly

$$V_B = \frac{Bpd}{\left[\ell\text{n}\left(\frac{Apd}{\ell\text{n}\left(1 + \frac{1}{\gamma}\right)}\right)\right]} \tag{2.7}$$

for molecular gas.

γ is actually also a function of E/p, similar to α. It has been found experimentally that γ is in fact fairly constant for most discharge conditions commonly encountered. Hence it can be seen that the breakdown voltage is a function of (pd) for both atomic and molecular gases. This has been demonstrated experimentally. From experiment, a minimum V_B is observed for a fixed value of (pd) for a certain discharge condition (gas type and electrode material). This minimum breakdown can be derived by differentiating (2.6) or (2.7) with respect to (pd) to be

$$(V_B)_{\min} = \frac{D^2}{4C}\left[7.39\ell\text{n}\left(1 + \frac{1}{\gamma}\right)\right] \quad \text{at} \quad (pd)_{\min} = \frac{7.39\ell\text{n}\left(1 + \frac{1}{\gamma}\right)}{C} \tag{2.8}$$

for atomic gas

$$(V_B)_{\min} = \frac{B}{A}\left[2.72\ell\text{n}\left(1 + \frac{1}{\gamma}\right)\right] \quad \text{at} \quad (pd)_{\min} = \frac{2.72\ell\text{n}\left(1 + \frac{1}{\gamma}\right)}{A}. \tag{2.9}$$

for molecular gas

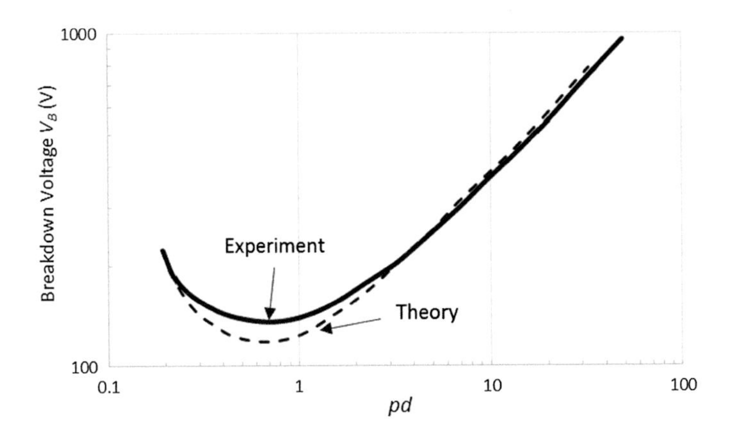

Fig. 2.5 The Paschen curve for air

The plot of breakdown voltage V_B against *pd* is often referred to as the Paschen Curve or the Paschen Law. The example for air is shown in Fig. 2.5.

It can be seen from the Paschen Curve that on the left hand side of the V_B minimum, the effect of increasing *pd* is to lower the breakdown voltage, while on the right hand side of the V_B minimum, the trend is just opposite. Increasing *pd* will now makes it more and more difficult for breakdown to occur.

2.1.2 *The* I–V *Characteristic of Electrical Discharge*

The variation of the current flowing in the electrical discharge circuit shown in Fig. 2.1 with applied voltage V_s can be summarized by the *I–V* characteristic curve as shown in Fig. 2.6. In this curve, the vertical axis is the voltage drop across the discharge tube.

The first part of the characteristic is caused by charges produced by background ionization of the gas either by environmental stray radiation or by photoelectric effect at the cathode surface due to UV radiation. At low voltage, whatever electrons available may be accelerated towards the anode to constitute to the current. If no ionizing collision by electron can occur due to the low potential (and hence low electric field), the maximum current that can be obtained is determined by the total number of initial electrons available. This current is in the region below nano-ampere and it increases with applied potential. It reaches a saturation value corresponding to the maximum number of electrons available.

With increasing applied potential, the electrons may be accelerated to energy above the excitation and ionization thresholds and these processes will then take place. New charge particles, both ions and electrons, will be produced by ionization and this gives rise to an increase in the discharge current. Eventually as the potential is further increased to reach the breakdown voltage, the discharge current will

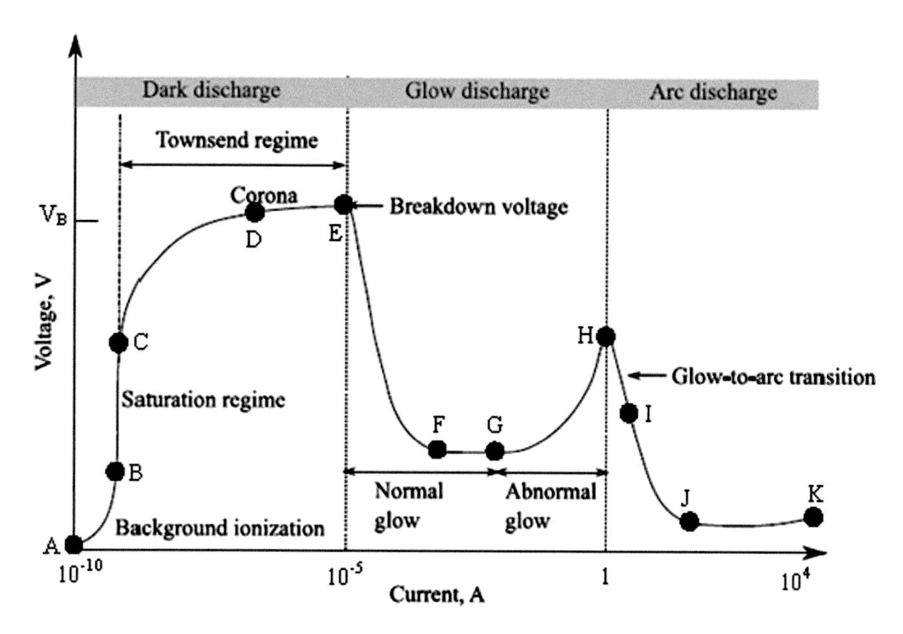

Fig. 2.6 *I–V* characteristic of electrical discharge

increase exponentially and then breakdown will occur and an electrical discharge is formed. This electrical breakdown will occur at potential V_B given by (2.6) or (2.7).

The region of the *I–V* curve before breakdown is often referred to as the dark discharge region. It is subdivided into the background current region, the Townsend region, and the corona region. A corona discharge is maintained by controlling the current at the micro amperes region, either by external or internal means. We will look into this again later. The voltage drop across the discharge tube when the discharge is in the dark discharge region is roughly equal to the applied potential.

After breakdown, the discharge will try to draw infinite current from the power supply so it is essential to have a current limiting resistor R_L in series between the source and the discharge. The type of discharge that is obtained will depend on the magnitude of the discharge current, which is controlled by the combined effect of the limiting resistor R_L and the plasma impedance. Ideally, the plasma resistance is negligible compared to R_L after breakdown. This means that the voltage drop across the discharge tube will be zero and the full voltage will be developed across R_L. However, when R_L is adjusted to limit the current to be in the region of mA, there will be a voltage drop across the discharge tube which is roughly constant when the discharge current is varied. This is the normal glow discharge region and the voltage across the electrodes is called glow voltage, V_g. The normal glow region may be extended down to 10^{-5} A when the current is reduced gradually from mA.

On the other hand, when the current is further increased to beyond 100 mA, the voltage across the electrodes will not remain constant but will increase. The glow discharge is said to become abnormal. When the current is increased to greater than

1 A, the voltage across the electrodes suddenly drops to lower than the glow voltage and the discharge has changed into the arc discharge. As a summary, the three types of discharge that can be obtained by controlling the current are:

$$10^{-7}\text{–}10^{-5}\ \mathrm{A} \Rightarrow \text{Corona discharge}$$
$$10^{-5}\text{–}1\ \mathrm{A} \Rightarrow \text{Glow discharge}$$
$$> 1\ \mathrm{A} \Rightarrow \text{Arc discharge}$$

2.1.3 The Corona Discharge

After breakdown, if the discharge current can be controlled at the level of several μA, a corona discharge will be obtained. The potential drop across the electrodes is still the same as the applied voltage. Corona discharge can also be obtained in situation where electrical breakdown voltage has not been reached but the electric field between the electrodes is not uniform. A particular situation is when the high voltage electrode (can be either anode or cathode) has a sharp profile, such as in the form of a needle or thin wire. In this case the electric field at the sharp point is sufficiently high (>30 kV/cm) that the electrons may be accelerated to high enough energy to produce ionizing collision leading to breakdown within a close distance from the sharp point. The distance from the sharp point within which electrical breakdown can occur is called the effective distance of corona discharge.

A phenomenological situation of corona discharge is illustrated in Fig. 2.7. With such a configuration, the Poisson Equation in spherical coordinate is written as:

$$\nabla \cdot E = \frac{1}{r^2}\frac{d}{dr}\left(r^2 E\right) = -\frac{\rho}{\varepsilon_o} \approx 0 \tag{2.10}$$

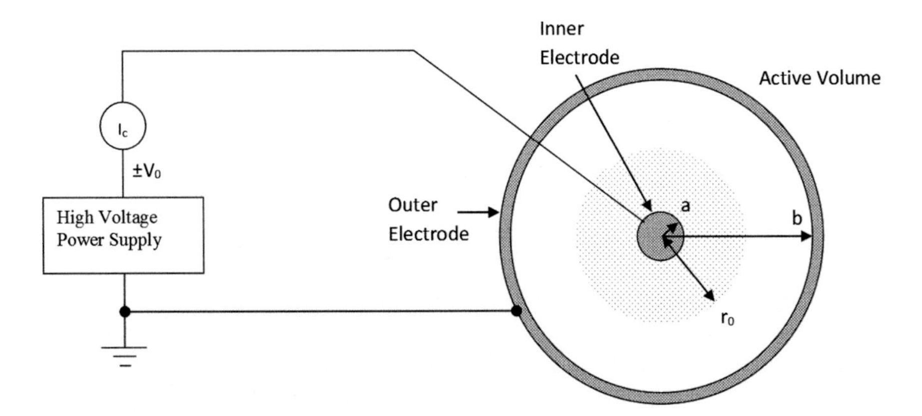

Fig. 2.7 Corona discharge at a sharp point

which gives the radial profile for the electric field between the electrodes expressed as:

$$E(r) = -\frac{dV}{dr} = \frac{a^2 E_o}{r^2}. \tag{2.11}$$

Integrating with the boundary conditions that at $r = a$, $V = V_o$ and at $r = b$, $V = 0$; we get,

$$V(r) = V_o \left[\frac{a(b-r)}{r(b-a)} \right]. \tag{2.12}$$

Hence,

$$E(r) = \frac{abV_o}{r^2(b-a)} \approx \frac{aV_o}{r^2} \quad \text{when } a \ll b. \tag{2.13}$$

At the surface of the sharp point with radius a, the electric field is $E_o \approx \frac{V_o}{a}$ which can be significantly large for small a. This may also be true for small r.

2.1.4 *The Glow Discharge*

The glow discharge is one of the most commonly used plasma in industry. Although it can be produced by a large variety of discharge configurations: DC, RF, DC or RF magnetron, ECR microwave discharge etc., the basic properties of the plasmas produced in these discharges are similar. In the classical configuration of the plane parallel electrodes placed inside a cylindrical glass chamber as shown in Fig. 2.8, operated in the normal glow discharge mode, the discharge consists of several bright and dark regions. The most prominent part of the discharge, which is the plasma proper by definition (zero or at least low electric field, charge neutrality etc.), is the positive column. In general, the electron temperature of this plasma is in the range of 1–2 eV while the ions and atoms are near to room temperature, and the electron density in the range of 10^6–10^8 cm^{-3} in the DC discharge case. With such conditions, the majority of the species present is in the neutral state, perhaps largely excited. A small fraction of singly ionized ions, and even smaller fraction of doubly ionized ions may also be present. According to Scottky's Diffusion Model where the production of new electrons by ionization is balanced by the loss of charge particles due to radial diffusion, the radial electron density profile in the positive column is expressed in terms of the zeroth order Bessel Function

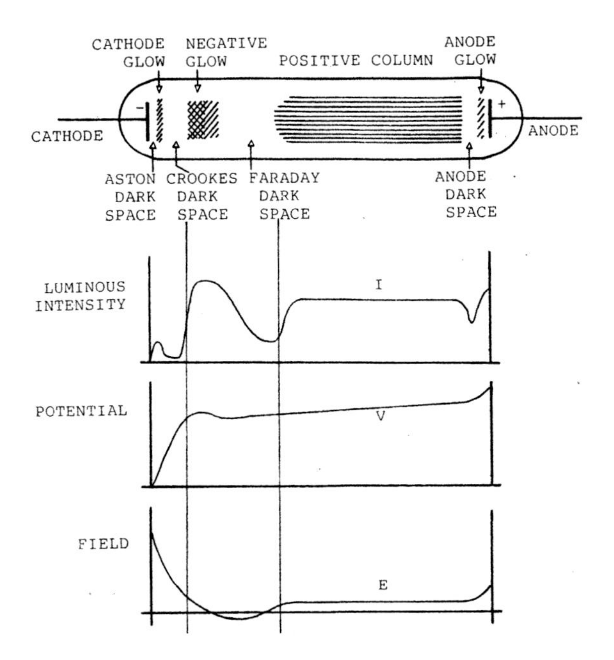

Fig. 2.8 Characteristic features of a normal glow discharge

$$n(r) = n_o J_o\left(2.405\frac{r}{R}\right), \tag{2.14}$$

where n_o is the electron number density at the axis of the column and R its radius.

Another region of interest is the cathode fall region consisting of the Aston Dark Space, Cathode Glow and the Crookes Dark Space. The potential drop sharply across this region resulting in high electric field. The high electric field is the main cause of energetic ionic bombardment at the surface of the cathode that results in electron emission. This is in fact an essential feature of the discharge that makes it "self-sustained". The cathode fall region is also the region where the electrons gain most of their kinetic energies. The thickness of the cathode fall region, d_c is relatively small compared to the length of the whole discharge tube. The potential drop across this region, on the other hand, can be shown to be given be

$$V_c = \frac{B}{A}\left[3\ell\text{n}\left(1 + \frac{1}{\gamma}\right)\right] \text{ for a molecular gas.} \tag{2.15}$$

This can be compared with the expression for the minimum breakdown voltage for a discharge in molecular gas given by Eq. (2.9). This means that the potential drop across the cathode fall region for a gas discharge is expected to be about the same as the minimum breakdown voltage for the setup, or vice versa. Similarly, for an monoatomic gas,

$$V_c \approx \frac{D^2}{4C}\left[7.39\ell\mathrm{n}\left(1+\frac{1}{\gamma}\right)\right]. \tag{2.16}$$

The other two regions, the negative glow and the Faraday Dark Space, are regions where the energetic electrons are thermalized before reaching the positive column. At the anode side, electrons near to the anode may experience slight increase in acceleration and bombard at the anode surface to produce a layer of bright region, the anode glow. Separating the positive column and the anode glow is the anode dark space.

2.1.5 *Hot Cathode Discharge*

In the glow discharge described in Sect. 2.1.4, no external means of heating is applied to the cathode. It is thus often referred to as a "cold cathode" discharge, although the electrodes do become hot, or at least warm during the discharge. We say it is cold is to contrast it with the case where the cathode is heated to high temperature when it starts to emit electrons. A tungsten filament is commonly used as the cathode in this case. Since we now have abundant of electrons emitted from the cathode, the bombardment of the cathode by energetic ions is no more required to keep the discharge self-sustained. In fact, this should be avoided since it will shorten the life of the filament. The schematic setup is as shown in Fig. 2.9.

Since there is now abundant supply of electrons from the hot cathode, a flow of current can be obtained even if the space between the electrodes is kept at absolute vacuum.

In theory, the emission of electrons from the filament is governed by its temperature T and the work function ϕ of the material used as the filament. This is given by the Richardson-Dushman equation:

$$J = AT^2 \exp\left(-\frac{e\phi}{kT}\right), \tag{2.17}$$

where A is a constant.

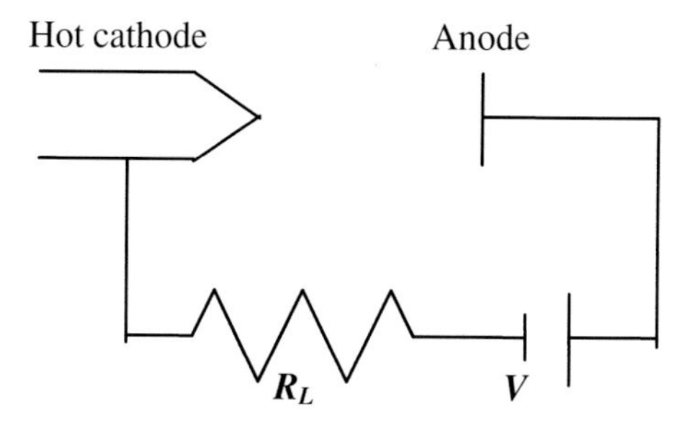

Fig. 2.9 Schematic of a hot cathode discharge

However, this current density cannot be achieved since the electrons emitted are not moved to the anode instantaneously and there will be a pile-up of electrons near the cathode. At equilibrium, a layer of space charge will be formed in front of the cathode. The distribution potential from the cathode to anode is such that almost all the potential drop is across this layer of space charge and the potential at the edge of the space charge is the same as that at the anode. The edge of the space charge layer thus acts as the virtual anode. The current density is given by

$$J = \frac{4\varepsilon_o}{9}\sqrt{\frac{2e}{m_e}}\frac{V^{3/2}}{d^2}, \tag{2.18}$$

where d is the inter-electrode spacing and V is the applied potential. This is, however, subjected to a maximum value determined by the Richardson-Dushman equation given by (2.17).

Note that the expression for J above applied to the vacuum case where the electrons are assumed to encounter no collisions on their ways to the anode. In the case when the inter-electrode space is filled with a gas at a fixed pressure, collision cannot be neglected and electron mobility through the gas should be considered. The discharge is said to be mobility limited and the discharge current density is given by

$$J = \frac{9\varepsilon_o\mu_e}{8}\frac{V^2}{d^3}, \tag{2.19}$$

where μ_e is the electron mobility.

2.1.6 The Arc Discharge

We have seen that when the current of the glow discharge is increased to above 1 A, the discharge will be transformed into the arc discharge. The potential drop across the electrodes will drop to a value less than the glow voltage. At low pressure, the discharge can be highly unstable and it may appear as a glow discharge with intermittent arcing.

(For practical reason, the transition of glow to arc discharge is seldom observed experimentally. The reason is that the DC power source used for the glow discharge is normally rated at a few kV and below 500 mA. To power the arc discharge, the required current rating is in the >A region. On the other hand, once a stable arc discharge is formed, only a relatively low voltage is required to maintain it. Hence, the power supply used for arc discharge is usually in the range of say 100 V, 100 A rating. The main factor here is, of course, the cost of the equipment.)

The arc discharge is usually operated at higher pressure than that of a glow discharge. The most common applications of the arc discharge in industry is to use it as a high temperature heat source, for example in plasma furnace. For these applications the arc discharge is operated at pressure above 10 torr and it is called

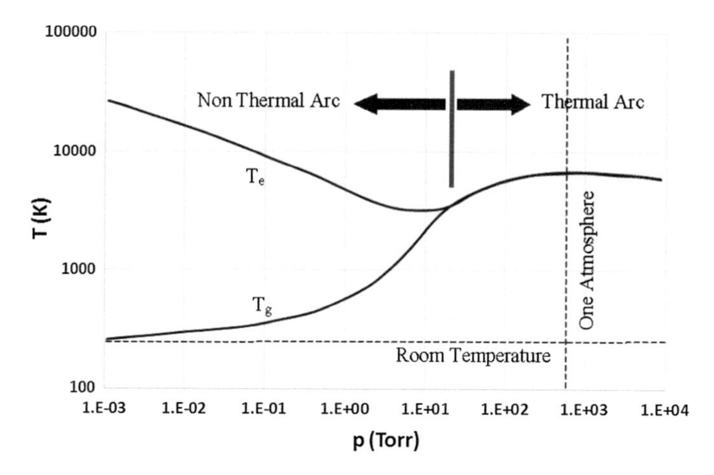

Fig. 2.10 The two regimes of arc discharge—thermal and non thermal

the thermal arc. At high pressure, the rate of collision is sufficiently high that the electrons and ions/atoms are able to reach thermal equilibrium that they will achieve a single temperature. For lower pressure operation, the electrons are at thermal equilibrium corresponding to a temperature higher than that of the ions/atoms. The two regimes of arc discharge are shown in Fig. 2.10.

Due to the high discharge current employed in the arc discharge, the plasma column exhibits two distinct features as compared to the glow discharge. The first is that the high discharge current causes a constriction effect on the plasma column. Secondly, the electrodes are severely heated and they should be cooled by running water. The heating effect will eject some of the electrode materials into the plasma and thus cools the plasma close to the electrode surfaces. As a consequence of this cooling effect this part of the plasma column is constricted to smaller radius than the rest of the column.

An important property of the thermal arc discharge operating at high pressure is its ability to achieve thermodynamic equilibrium. This qualify the plasma formed to act as a blackbody radiator. The radiation spectrum is expressed by Planck Law as

$$\frac{dM_E}{d\lambda} = \frac{2\pi hc^2}{\lambda^5}\frac{1}{\exp(hc/kT\lambda) - 1}\ \left(\mathrm{W/m^2/m}\right) \tag{2.20}$$

This is plotted as shown in Fig. 2.11.

The peak emission is at a wavelength of

$$\lambda_{MAX} \approx \frac{hc}{5kT}(\mathrm{m}) \approx \frac{2500}{kT(in\,\mathrm{eV})}\ (\mathrm{\AA}) \tag{2.21}$$

Thus for arc discharge at temperature of 1 eV, the peak of the blackbody radiation is in the UV region. This may give rise to occupational hazard in using the arc discharge heat source.

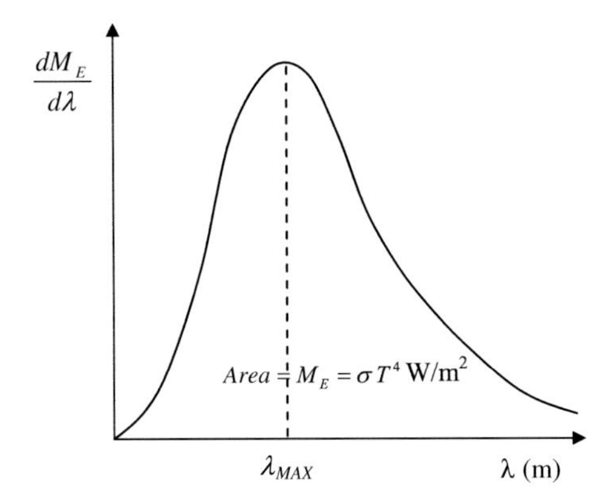

Fig. 2.11 The emission spectrum of a black body radiator

For application purposes, the arc discharge is normally operated in two configurations: the plasma torch and the plasma spray. The plasma torch is used in applications such as the plasma furnace for solid waste disposal. There are two modes of plasma torch, one based on the transferred arc concept and another based on the non-transferred arc concept. These are shown in Fig. 2.12. In the transferred arc mode, the work-piece becomes one of the electrodes, normally connected to earth potential. This will work if the work-piece is a conducting material. This mode of operation is used for metal cutting or melting. In the non-transferred mode, the arc discharge is formed between the cathode and the anode (ground) which is in the form of nozzle. The working gas is flown through the nozzle so that the plasma produced is in the form of a jet of high temperature. This can be used to treat insulating matcrials as well.

Since the thermal arc discharge plasma is said to have achieved thermodynamic equilibrium, the gas temperature (ions and atoms) is expected to be equal to that of the electrons. A temperature of up to 20,000 K (equivalent to 2 eV) can be expected.

The plasma spray is a variation of the non-transferred mode plasma torch. It can be used to spray coat material (for example ceramics) onto a substrate, or to melt

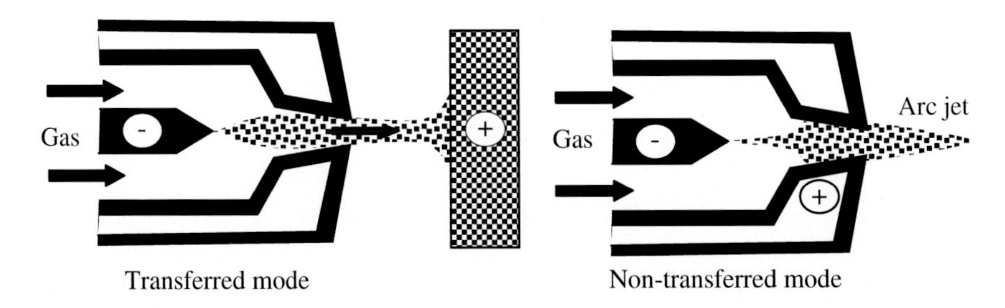

Fig. 2.12 Two possible modes of arc discharge configuration

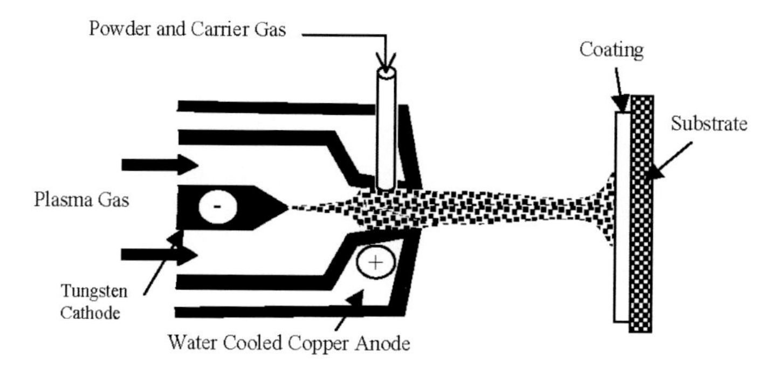

Fig. 2.13 Plasma spraying using the arc discharge

and rapidly solidify heavy metals such as molybdenum or tungsten or their composites to form power. An example of the setup for plasma spray coating of material is shown in Fig. 2.13.

2.2 AC (Radiofrequency) Discharge

Gas discharge powered by an AC source can be considered not different from that maintained by a DC source if the frequency is low. The criterion of "low" frequency is that the characteristic time of the voltage variation (usually taken to be the periodic time) should be much larger than the transit time of the ions from anode to cathode. This is normally in the region of below 1 kHz. For high frequency, the discharge behavior becomes different from the DC case. First, the breakdown voltage will be lower. In a self-sustained glow discharge, new charged particles are produced by the ionization of the gas by electron collisions and the secondary emission at the cathode surface by ion bombardment and this is balanced by the loss of electrons at the anode. We see from Eq. (2.5) that breakdown will occur when this balance is achieved. For sufficiently high frequency (probably above 1 MHz) source, the loss of electrons at the anode will be reduced (or even becomes zero loss) since the alternating field may reverse some or all of the electrons before they reach the electrodes. Although there will also be a reduction of electron production by ion bombardment at the cathode surface, this will be compensated by the reduction of loss of electrons at the anode and also the increase in ionization in the gas since the electrons now remain in the plasma for longer time as they are moving back and forth following the electric field. In this case the main electron loss mechanism is radial diffusion.

As we can see from the above discussion, one favorable condition for AC powered gas discharge is to employ sufficiently high frequency to ensure that the oscillation time of the electrons caused by the alternating electric field is shorter than their transit time between the electrodes, or $\omega\tau < 1$ where ω is the angular frequency of the field and τ is the electron transit time between electrodes. This is also dependent on the pressure of the gas as at lower pressure the electron transit time is expected to decrease. For pressure of a few torr and discharge distance of several cm, the use of frequency in the radiofrequency range is appropriate. A commonly used frequency is 13.56 MHz as agreed by international communication authorities. Under this condition we expect $\omega < \upsilon$ where υ is the electron collision frequency. This means that an electron collides many times within one oscillation of the electric field so it may be able to transfer the energy it absorbs from the field to other particles. For lower pressure, say in the region of mtorr, there may not be sufficient collision for the electrons to achieve equilibrium with other particles so that it is more effective to heat the electrons collectively. This requires frequency $\sim$ electron plasma frequency ω_{pe} which is at the microwave region (>GHz).

In an RF discharge, the plasma will not be "off" between cycles of the electric field variation which is normally sinusoidal, although the electrons are expected to travel back and forth between the electrodes as the field changes direction. As a result, a steady plasma is formed between the electrodes as seen in Fig. 2.14. The condition of the plasma is expected to be similar to that of the DC discharge except that the potential distribution between the electrodes may vary during each cycle.

An RF discharge can be produced by using 2 types of configuration: capacitively coupled or inductively coupled. The capacitive coupling can be implemented with a set of parallel plates such as those used for the conventional DC glow discharge shown in Fig. 2.14. For an RF power source, an electrode less discharge can be obtained by placing the electrodes outside the plasma chamber (Fig. 2.15). This can eliminate the contamination of the plasma by the electrode materials.

Electrode less discharge can also be produced by using the inductively coupled configuration. An example of this configuration is to wrap a solenoid of N turns,

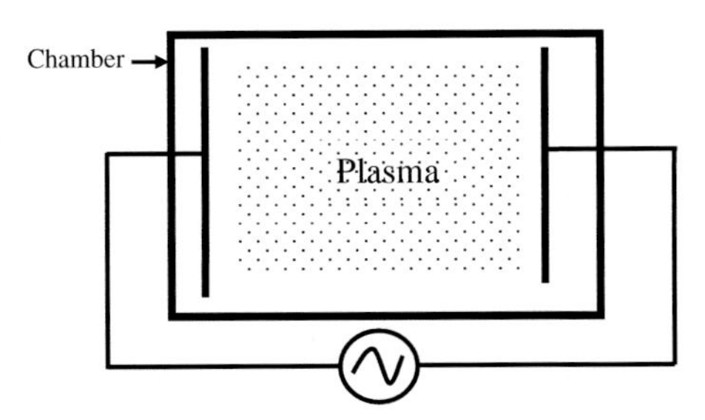

Fig. 2.14 Schematic of an AC glow discharge

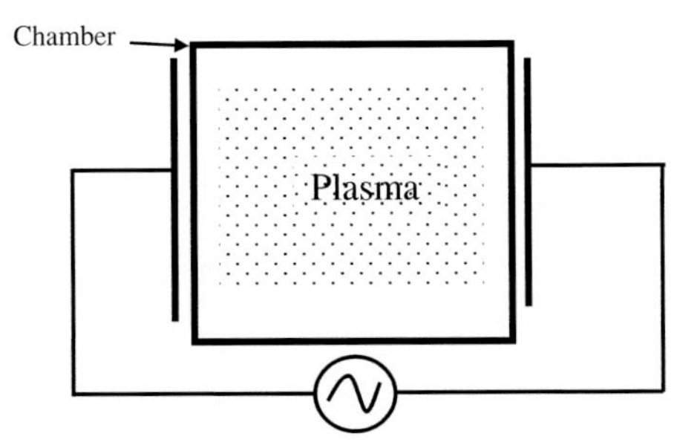

Fig. 2.15 The inductive mode of AC glow discharge

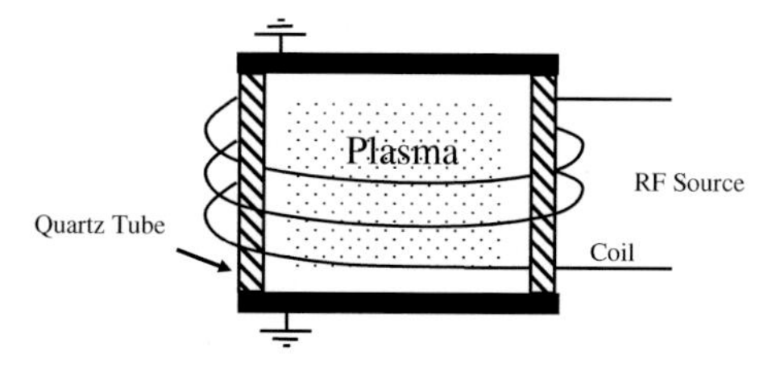

Fig. 2.16 Inductively coupled plasma using selonoid

length L and radius r around a glass tube (Fig. 2.16) and a discharge inside the tube can be produced by the electric field induced by the high frequency current I flowing through the solenoid.

In this case, there are two components of the induced electric field, one along the axis of the solenoid (E_z) and one in the azimuthal direction (E_θ) given by:

$$E_z = \frac{\mu_o \omega N^2 \pi r^2 I}{L^2} \quad \text{and} \quad E_\theta = \frac{\mu_o \omega N r I}{2L}$$

Under normal condition, $E_z > E_\theta$ until breakdown in the azimuthal direction occurs and E_z will be reduced by the field induced by the toroidal plasma current.

Another possible form of inductively coupled plasma discharge system is by using a planar coil instead of the cylindrical solenoid. The planar coil is placed outside the discharge chamber, separated by a dielectric window usually made of quartz. An example of such a system is as shown in Fig. 2.17.

In this configuration, we now have $E_\theta(r, z)$ instead of E_z together with E_θ. The expressions for the capacitive field and the toroidal field are complicated and will not be discussed here. However, the principle of operation in the two cases (cylindrical solenoid and planar) are identical. As the input RF power is increased gradually, the discharge is expected to breakdown due to the capacitive field produced by the coil. Hence at low RF power, the discharge is in the direction of the capacitive field and it is said to be operated in E-mode. When the power is raised to

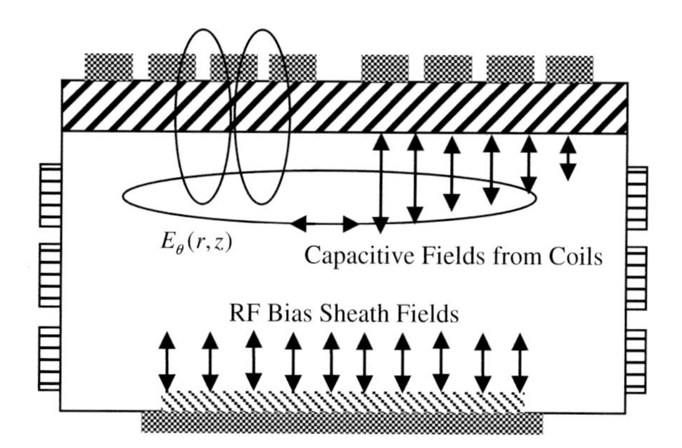

Fig. 2.17 Planar coil inductively coupled plasma

sufficiently high level, breakdown in the azimuthal direction will occur. The azimuthal discharge is able to draw higher current than that in the axial discharge because it is acting as the single loop secondary of the transformer with the solenoid coil as the primary. This is the H-mode operation. The E to H mode transition is usually abrupt and can be observed distinctly. The plasma produced has density several order of magnitude higher than that produced in the E-mode operation. This can be observed experimentally from the spectroscopic measurement of a designated line radiation of an ionic specie. An example is shown in Fig. 2.18 [2] which shows the jump in the intensity of the 394.6 nm line of Ar^+ in an argon discharge. The abrupt increase in the intensity of this line is caused by an abrupt increase in the number density of the Ar^+ specie which may be due to a sharp increase in the electron temperature of the plasma during the E to H mode transition.

A crucial problem in the design of an RF discharge circuit is the efficient coupling of the RF power source to the plasma. This can be achieved by connecting the RF power source to the load (parallel plate electrodes as capacitive load; cylindrical solenoid or planar coil as inductive load) through a matching network. In general, the plasma load can be considered to be a combination of inductance L, capacitance C and resistance R. In an RF circuit, we can define the admittance of the load to be $Y = G + jB$, where

$$G = \frac{R}{R^2 + (X_C + X_L)^2}$$

is the conductance, with $X_C = \frac{1}{\omega C}$ and $X_L = \omega C$,

$$B = \frac{X_C + X_L}{R^2 + (X_C + X_L)^2}$$

is the susceptance.

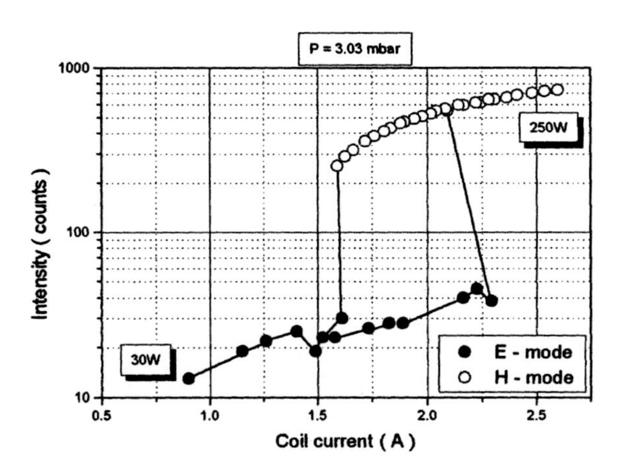

Fig. 2.18 The sharp increase spectral line intensity during E-H mode transition

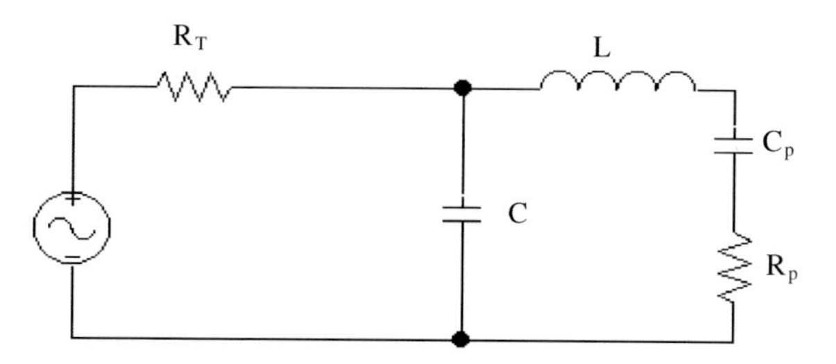

Fig. 2.19 Matching network for a capacitively coupled discharge

This load can be matched by an RF source with $R_T = \frac{1}{G}$ and $C_T = -\frac{B}{\omega}$. For a capacitively coupled discharge, the load can be represented by a capacitor and a resistor in series as shown in Fig. 2.19. Matching can be achieved with

$$C = \frac{1}{\omega}\left(\frac{1}{R_T R_p} - \frac{1}{R_T^2}\right)^{\frac{1}{2}} \quad \text{and} \quad L = \frac{1}{\omega}\left[(R_p R_T - R_p^2)^{\frac{1}{2}} + \frac{1}{\omega C_p}\right]$$

For an inductively coupled plasma, it can be represented by an inductance and a resistance in series as shown in Fig. 2.20.

For this case, matching is achieved when

$$C_2 = \frac{1}{\omega}\left(\frac{1}{R_T R_p} - \frac{1}{R_T^2}\right)^{\frac{1}{2}} \quad \text{and} \quad C_1 = \frac{1}{\omega}\left[\frac{1}{\omega L_p - (R_p R_T - R_p^2)^{\frac{1}{2}}}\right]$$

Another type of matching network used for inductively coupled plasma is to use only C_1 instead of both C_1 and C_2 as discussed above. In this case C and L_p forms an LC resonant circuit which will achieve maximum current flow when $\omega L_p = \frac{1}{\omega C}$. For this case, $B = 0$ and $Y = G = 1/R_p$. We require $R_T = R_p$ for the circuit to match. However, since R_p is usually varying with plasma condition this is difficult to achieve. For discharge in E mode, R_p is normally negligibly low so matching has to be done by balancing X_L and X_C (varying C) so that the net impedance is equal to R_T.

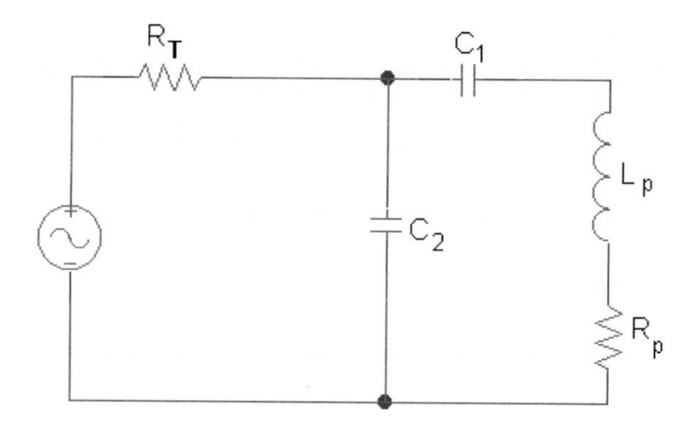

Fig. 2.20 Matching network for an inductively coupled discharge

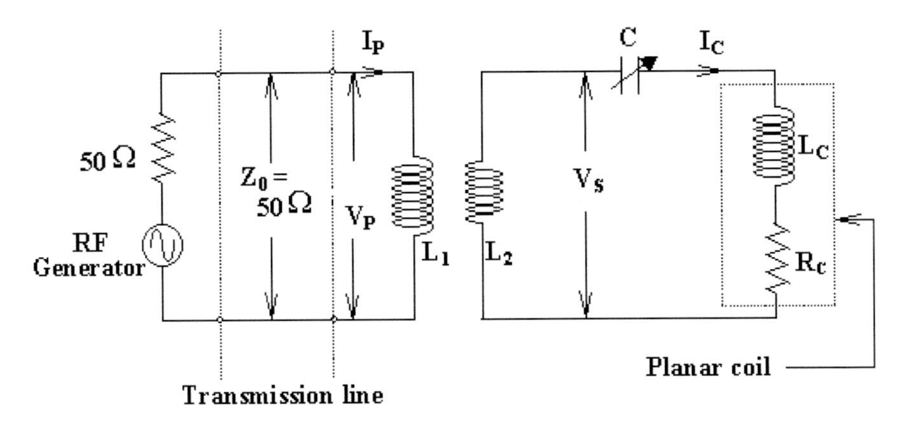

Fig. 2.21 Matching network of an inductively coupled discharge using resonant circuit

This requires R_T to be as low as possible if we want to produce discharge current of as near to the peak of the resonance as possible. The commonlly available RF power source with 50 Ω output impedance is not suitable in this case. The source impedance suitable for this purpose should be as low as possible, may be 1 Ω or lower. One possible solution to this problem is to add a step down transformer between the RF generator and the *LC* circuit to provide an effective low output impedance for the generator. This arrangement works well for E mode discharge. However, for the H mode discharge, R_p may be higher so the matching has now to be achieved by keeping $\omega L_p = \frac{1}{\omega C}$ while the impedance of the source is adjusted to be equal to R_p. This is difficult to achieve and the circuit in Fig. 2.21 may be more suitable.

Plasma torch powered by RF power source can also be built using the inductively coupled configuration. The most common application of this plasma source is for elemental analysis by spectroscopic method. The equipment used for this purpose is normally called the "ICP" which actually means inductively coupled plasma. The plasma is produced by heating the stream of gas (at atmospheric pressure) flowing through the inside of a solenoid coil. The sample to be analyzed may be in the gas form or for liquid sample, it will be heated to form vapor to pass through the torch.

2.3 Microwave Heating of Plasma

The microwave is the region of electromagnetic spectrum with frequency above 1 GHz up to 300 GHz corresponding to wavelength of 30 cm to 1 mm respectively. This is the range of electron plasma frequency for plasma with electron density in the range of 1×10^{10} cm^{-3} to 1×10^{15} cm^{-3}. Hence when the microwave is used for heating a plasma at such a range of density, the microwave energy will be absorbed by the electrons collectively and not individually. In contrast to individual particle heating (such as DC or RF discharge), the electrons gain energy by absorbing the microwave energy directly and convert it into kinetic energy.

Assuming that the microwave can be described by the wave equation:

$$\mathbf{E} = \mathbf{E_o} \exp[i(\kappa.\mathbf{r} - \omega t)] \tag{2.22}$$

where $\boldsymbol{E}$ is the electric field with magnitude E_o, κ is the wave vector, $\boldsymbol{r}$ is the position vector, and ω is the angular frequency of the wave. The rate of energy absorption by electrons can be shown to be given by:

$$P = \frac{n_e \cdot \mathrm{e}^2 \cdot E_o^2}{2m_e \cdot \nu} \cdot \left(\frac{1}{1 + (\frac{\omega}{\nu})^2} \right) \tag{2.23}$$

where ν is the electron-atom/ion collision frequency, n_e is the electron number density and m_e is the electron mass. It can be seen that P is maximum when $\omega \ll \nu$. This requires that the plasma density should not be too low. A schematic of the typical setup used for this method of plasma heating is shown in Fig. 2.22.

The microwave power can be coupled to the gas inside the chamber by either the resonant cavity configuration or the multimode cavity configuration. In the resonant cavity configuration, the dimensions of the cavity, R and d are chosen to be equal to the wavelength of the microwave, that is $R = \lambda$, and $d = \lambda$, where $\lambda = 2\pi c/\omega$. In the multimode configuration, the conditions are $R > \lambda$, and $d > \lambda$. The plasma produced by the multimode configuration is expected to be more uniform.

For low plasma density where $\omega > \nu$, the electron heating by microwave can be enhanced by applying an external magnetic field. For the case where this external magnetic field is in the direction parallel to the direction of propagation of the wave, the wave will become circularly polarized, in both Right and Left directions. The rate of wave energy absorption by the electron can now be expressed in 2 components as: $P = P_R + P_L$ where

$$P_R = \frac{n_e \cdot \mathrm{e}^2 \cdot E_o^2}{2m_e \cdot \nu} \cdot \frac{1}{2} \left(\frac{1}{1 + (\frac{\omega - \omega_{ce}}{\nu})^2} \right) \text{ (R wave)} \tag{2.24}$$

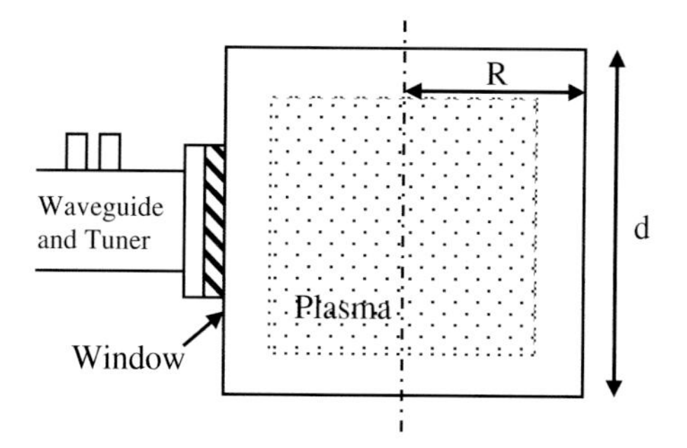

Fig. 2.22 Schematic of a simple system of microwave heating of plasma

$$P_L = \frac{n_e \cdot e^2 \cdot E_o^2}{2m_e \cdot \nu} \cdot \frac{1}{2}\left(\frac{1}{1+\left(\frac{\omega+\omega_{ce}}{\nu}\right)^2}\right) \text{ (L wave)} \tag{2.25}$$

Here $\omega_{ce} = \frac{e \cdot B}{m_e}$ is the electron cyclotron frequency of the magnetized plasma. Hence it can be seen that the absorption of microwave energy by electron will be a maximum when the frequency of the wave is the same as the electron cyclotron frequency, $\omega = \omega_{ce}$. This is known as *Electron Cyclotron Resonant (ECR) heating.*

Since the magnetic field distribution is not uniform throughout the volume inside the chamber, the condition of $\omega = \omega_{ce}$ may only be met in certain area and this is called the resonant surface. It is within this resonant surface that intense plasma heating will occur. The electron density produced by the ECR heating can be easily more than an order of magnitude higher than that achieved in the normal microwave plasma heating. Electron temperature of 5–10 eV can be achieved, as compared to 1–2 eV in most other DC or RF discharges.

2.4 Pulsed Plasma Discharges

The plasmas produced by DC or RF discharge or by microwave heating are at steady state. The plasma electron temperature of these plasmas is below 10 eV and the electron density is below 10^{15} cm^{-3}. In order to produce hotter and denser plasma, a power source capable of generating higher power density will be required. This can be achieved by using a capacitor discharge where the energy is first stored in the capacitor and then discharge through the gas to produce the plasma. Power density of up to 10^{18} W/m^3 or higher can be generated easily. The plasma produced is able to achieve condition near to the fusion plasma condition.

The simplest technology used to power a pulsed plasma discharge is the capacitor discharge system. In this case a capacitor C is first charged up to a high voltage V so that an amount of energy of $\frac{1}{2}CV^2$ is stored in the capacitor. This energy is transferred to the plasma load via a switch. The schematic circuit of this capacitor discharge system is shown in Fig. 2.23.

In this circuit the plasma is represented by a time varying inductance $L_p(t)$ and a resistance $R_p(t)$ in series. L_o and R_o are the stray inductance and resistance of the circuit. Stray inductance and resistance may be contributed by the cables used to connect the various components together as well as contributed by the switch. The capacitor or a bank of capacitors used also has an internal inductance. All these have been lumped together and represented by L_o and R_o. The circuit represents a typical *LCR* circuit of which the circuit equation can be written as:

$$V_o = \frac{d}{dt}(LI) + IR + \frac{\int I\,dt}{C} = \left[L_o\frac{dI}{dt} + IR_o + \frac{\int I\,dt}{C}\right] + V_p \tag{2.26}$$

Fig. 2.23 Schematic of a capacitor discharge circuit

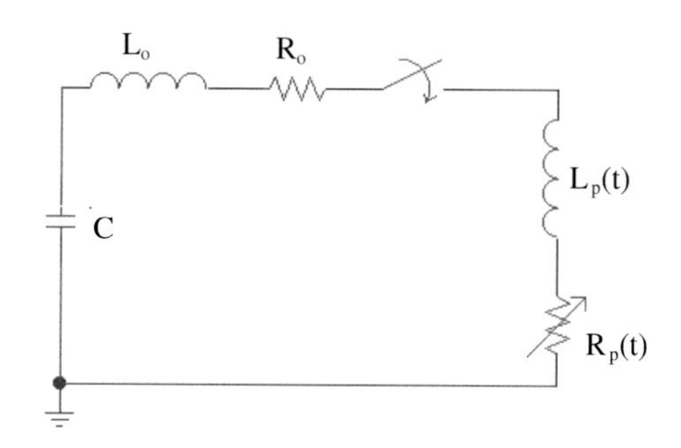

where

$$L = L_o + L_p, \quad R = R_o + R_p$$

and

$$V_p = \frac{d}{dt}(L_p I) + I R_p$$

is the voltage across the plasma.

Equation (2.26) can be rewritten in the form

$$\frac{d^2 I}{dt^2} + \frac{R}{L}\frac{dI}{dt} + \frac{1}{LC} = 0$$

which has a general solution of the form:

$$I(t) = [A\exp(nt) + B\exp(-nt)]\exp\left(-\frac{R}{2L}t\right), \tag{2.27}$$

where

$$n = \sqrt{\left(\frac{R}{2L}\right)^2 - \frac{1}{LC}}$$

Define the scaling parameter

$$\alpha = R/\sqrt{\frac{L}{C}} = R\sqrt{\frac{C}{L}},$$

then

$$n = \sqrt{\omega^2\left[\left(\frac{\alpha}{2}\right)^2 - 1\right]}$$

We can see that n may be imaginary, zero or real corresponding to $\alpha < 2$, $\alpha = 0$ or $\alpha > 2$.

Depending on n, I(t) may take three distinct forms:

(a) n = *imaginary or* $\alpha < 2$. In this case the solution becomes

$$I(t) = \frac{V_o}{\omega L} \sin \omega t \exp\left(-\frac{R}{2L}t\right).$$

Here $\omega \approx \sqrt{(-1)} \times n$ for $\alpha < 2$. The current waveform is a damped sinusoid with angular frequency ω and damping time constant $2L/R$. The oscillation is said to be *under damped.*

(b) $n = 0$ *or* $\alpha = 2$. In this case the solution becomes

$$I(t) = \frac{V_o}{L} t \exp\left(-\frac{R}{2L}t\right).$$

This is a waveform which decays with a time constant $2L/R$, the shortest time possible for the current to decay to zero. The waveform is said to be *critically damped.*

(c) n = *real or* $\alpha > 2$. In this case the solution becomes

$$I(t) = \frac{V_o}{nL} \sinh(nt) \exp\left(-\frac{R}{2L}t\right).$$

The waveform now decays with a time constant $>2L/R$. It is said to be over damped.

In principle, it is ideal to make use of the critically damped discharge for heating plasma. However, the plasma impedance is time varying so it is difficult to maintain the matching condition for critically damped discharge. For a high temperature plasma, the plasma resistance is normally low so the plasma can be represented by an inductance only. In this case the discharge is expected to be under-damped. On the other hand, in a high pressure pulsed discharge, the plasma resistance is sufficiently high that it may be possible to achieve the condition for critically damped discharge. One such example is the high pressure flash lamp.

2.4.1 *Pulsed Arc Discharge in High Pressure Flash Lamp*

The high pressure flash lamp is a high power arc discharge that is commonly used as a light source. One such application is the use of linear flash lamp for the pumping of solid state laser (for example ruby laser) or dye laser.

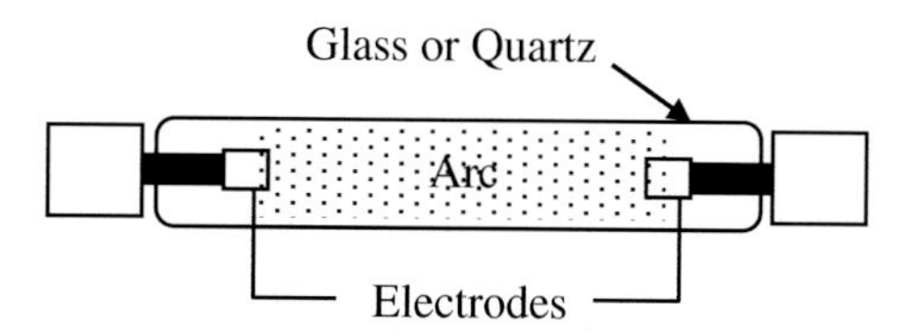

Fig. 2.24 Schematic of a linear flash lamp

An example of the linear flash lamp is as shown in Fig. 2.24.

The schematic circuit of the flash lamp discharge is similar to that shown in Fig. 2.23 except that the plasma is now represented by a fixed inductance (unknown) and a resistance (fixed or time varying).

Here we may combine the fixed inductance of the flash lamp and the stray inductance of the circuit into L and the stray resistance of the circuit is assumed to be negligible. Hence the circuit equation can be written as:

$$V_o = \frac{\int Idt}{C} + L\frac{dI}{dt} + V_p \tag{2.28}$$

where

$$L = L_o + L_p$$

which is constant.

And

$$V_p(t) = IR_p + L_p\frac{dI}{dt}.$$

Equation (2.28) can be normalized and solved for $I(t)$. This can be done by defining

$$\imath = \frac{I}{I_o}, \quad \tau = \frac{t}{t_o}$$

where

$$I_o = V_o\sqrt{\frac{C}{L}} \quad \text{and} \quad t_o = \sqrt{LC}$$

The normalized form of (2.28) is

$$\frac{d\imath}{d\tau} + \alpha\imath + \int \imath d\tau = 1 \tag{2.29}$$

where

$$\alpha = \frac{R_p}{Z_o} \quad \text{and} \quad Z_o = \sqrt{\frac{L}{C}}$$

This is actually the same as solving the ideal LCR discharge discussed earlier numerically and by choosing the value of α to be <2, =2 and >2 will give rise to solution for light damping, critical damping and over damping. The boundary conditions are: $\tau = 0$, $\iota = 0$, $\int \iota d\tau = 0$, $d\iota/d\tau = 1$. The solutions for the three cases are shown in Fig. 2.25.

It is clear that to obtain a single flash with optimum intensity and shortest pulse width, it is best to operate with critical damping condition. For an under-damp flash lamp discharge, the output light pulse may have a long tail and several humps in the light intensity may be observed as shown in Fig. 2.26 [3].

Note that in the above consideration, we have assumed that the resistance of the flash lamp is constant. In actual situation the resistance has been found to be dependent on the discharge current. It has been determined empirically that the voltage across the flash lamp can be expressed as $V_p = \pm k_o\sqrt{|I|}$, where k_o is often referred to as the flash lamp constant given by:

$$k_o = 1.28\left(\frac{\ell}{d}\right)\left(\frac{P}{X}\right)^{\frac{1}{5}} \tag{2.30}$$

where

X a constant of the gas (e.g. 450 for xenon, 805 for krypton as determined empirically),
ℓ the length of the flash lamp,
d the diameter of the flash lamp, and
P the pressure of the gas inside the flash lamp

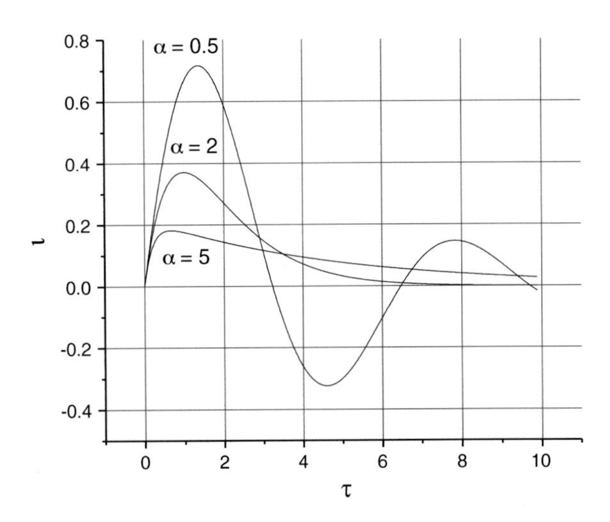

Fig. 2.25 Current waveforms for under-damped ($\alpha < 2$), critically damped ($\alpha = 2$) and over-damped ($\alpha > 2$) LCR discharge

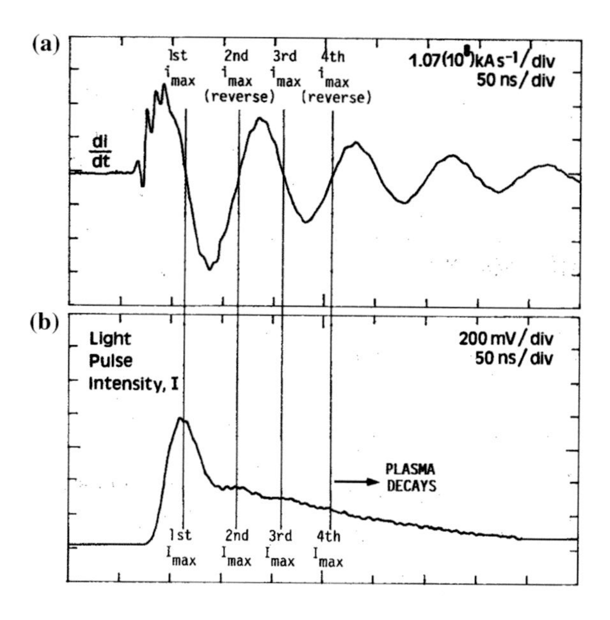

Fig. 2.26 Example of light output from a flash lamp powered by under-damped LCR discharge (Reprinted from Ref. [3], Copyright (1989), with permission from AIP)

The circuit equation is then written as

$$V_o = \frac{\int I dt}{C} + L_o \frac{dI}{dt} \pm k_o \sqrt{|I|} \tag{2.31}$$

and the normalized form becomes

$$\frac{d\iota}{d\tau} \pm \alpha |\iota|^{\frac{1}{2}} + \int \iota d\tau = 1 \; [\text{take} + \text{when } \iota = +, \text{take} - \text{when } \iota = -] \tag{2.32}$$

where $\alpha = \frac{k_o}{\sqrt{V_o Z_o}}$ is now the damping factor. An example of the solutions obtained for $\alpha = 0.2$, $\alpha = 0.8$ and $\alpha = 2$ are shown in Fig. 2.27. It can be seen that $\alpha = 0.8$ corresponds to the case of critical damping.

2.4.2 *Inductive Model of Pulsed Discharge—Shock Heating*

In the flash lamp discharge discussed above, the plasma is heated by the joule heating effect. In this case the discharge is described by the resistive model where the plasma is represented by a resistance and its inductance is either assumed to be constant or negligible. This is due to the fact that the plasma column produced by the discharge has a fixed geometry. In some cases such as the electromagnetic shock tube or the pinches, the current is arranged to flow in a thin sheet and it is driven by the self electromagnetic force (the $\boldsymbol{J} \wedge \boldsymbol{B}$ force) to supersonic speed so

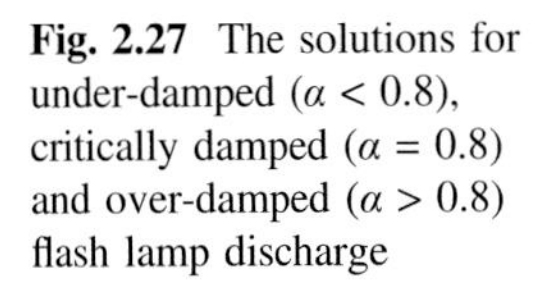

Fig. 2.27 The solutions for under-damped ($\alpha < 0.8$), critically damped ($\alpha = 0.8$) and over-damped ($\alpha > 0.8$) flash lamp discharge

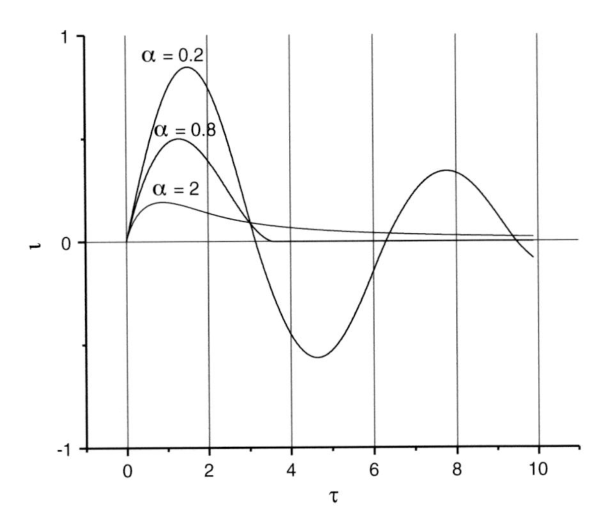

that shock heating of the plasma column can occur. The discharge is modeled by the inductive model where the plasma is represented by a time varying inductance. The inductance is time varying because its geometry is changing dynamically.

We consider two types of dynamics that may give rise to shock heating: the electromagnetic shock tube in which the current sheet is moving axially and the pinch (Z- and θ-pinches) where the current is moving radially towards the axis of the plasma column. Before we begin with these plasma devices we shall first discuss the shock heating effect.

2.4.2.1 Shock Heating of Plasma

Shock heating is a phenomenon where the particles in the plasma are driven to supersonic speed. In general, every medium has a characteristic speed by which information will propagate. An example is the speed of sound in air. If a piston is accelerated to a speed faster than the speed of sound, it is said to be supersonic and the perturbation created by the piston will pile up in front of the piston to form a shock front as illustrated in Fig. 2.28.

The shock front is the imaginary boundary formed in front of the supersonic piston which separates the ambient and the shock heated gas or plasma, Particles in front of the shock front are unperturbed while the particles after the shock front are being pushed by the piston to the speed of the piston and is said to have been shock-heated. The layer of shock heated gas in between the piston and the shock front will grow to be thicker and forms the plasma.

Consider the frame of reference of the shock front (that means you move with the shock front), the gas on the right hand side is at ambient while the gas on the left hand side has been shock heated.

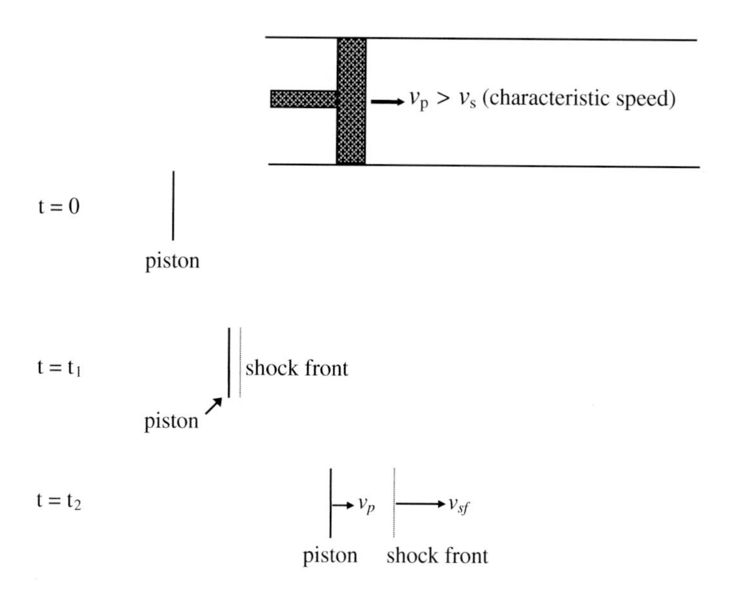

Fig. 2.28 Schematic illustration of the shock wave formation

$$\begin{array}{c|c} q_2 \quad \leftarrow & \leftarrow \quad q_1 \\ \mathrm{P}_2, \rho_2, T_2, h_2 & \mathrm{P}_1, \rho_1, T_1, h_1 \end{array}$$

where

$$q_1 = v_{sf}$$
$$q_2 = v_{sf} - v_p$$

For this system we can write:

$$\rho_2 q_2 = \rho_1 q_1 \text{ (conservation of mass across the shock front)} \tag{2.33}$$

$$\rho_2 q_2^2 + P_2 = \rho_1 q_1^2 + P_1 \text{(conservation of momentum across the shock front)} \tag{2.34}$$

$$\rho_2 q_2 \left(\frac{1}{2} q_2^2 + h_2 \right) = \rho_1 q_2 \left(\frac{1}{2} \rho_1^2 + h_1 \right) \text{ (conservation of energy)} \tag{2.35}$$

These three equations are called the shock jump equations. Another two relevant equations are:

$$P_2 = \rho_2 \frac{R_0}{M} T_2 (1 + \alpha_1 + 2\alpha_2 + \cdots) \text{ (equation of state)} \tag{2.36}$$

$$h_2 = \frac{\gamma}{\gamma - 1}\frac{P_2}{\rho_2} \text{ (enthalpy)} \tag{2.37}$$

It is useful to consider strong shock with the conditions that

$$P_2 \gg P_1$$
$$h_2 \gg h_1$$

Then Eqs. (2.34) and (2.35) will be reduced to

$$\rho_2 q_2^2 + P_2 = \rho_1 q_1^2 \tag{2.38}$$

$$\rho_2 q_2\left(\frac{1}{2}q_2^2 + h_2\right) = \rho_1 q_2\left(\frac{1}{2}\rho_1^2\right) \rightarrow \left(\frac{1}{2}q_2^2 + h_2\right) = \left(\frac{1}{2}\rho_1^2\right) \tag{2.39}$$

We now define the density ratio $\Gamma = \frac{\rho_2}{\rho_1} = \frac{q_1}{q_2}$ and substitute h_2 from (2.37) into (2.39),

$$\frac{1}{2}q_2^2 + \frac{\gamma}{\gamma - 1}\frac{P_2}{\rho_2} = \frac{1}{2}q_1^2$$
$$\therefore \quad \frac{P_2}{\rho_2} = (\frac{\gamma - 1}{\gamma})\frac{1}{2}(q_1^2 - q_2^2)$$
$$= \frac{1}{2}(\frac{\gamma - 1}{\gamma})q_1^2[1 - (\frac{q_2}{q_1})^2]$$
$$= \frac{1}{2}(\frac{\gamma - 1}{\gamma})q_1^2[1 - (\frac{1}{\Gamma})^2]$$
$$\frac{P_2}{\rho_2} = \frac{1}{2}(\frac{\gamma - 1}{\gamma})q_1^2(\frac{\Gamma^2 - 1}{\Gamma^2}) \tag{2.40}$$

Also, from (2.38),

$$P_2 = \rho_1 q_1^2 - \rho_2 q_2^2 = \rho_2 q_2^2[\frac{\rho_1 q_1^2}{\rho_2 q_2^2} - 1]$$
$$\therefore \quad \frac{P_2}{\rho_2} = q_2^2(\Gamma - 1) = q_1^2(\frac{q_2^2}{q_1^2})(\Gamma - 1) = q_1^2(\frac{\Gamma - 1}{\Gamma^2}) \tag{2.41}$$

Combining (2.40) and (2.41),

$$\frac{1}{2}(\frac{\gamma-1}{\gamma})(\frac{\Gamma^2-1}{\Gamma^2}) = \frac{\Gamma-1}{\Gamma^2}$$

$$\frac{1}{2}(\frac{\gamma-1}{\gamma})(\Gamma+1)(\Gamma-1) = (\Gamma-1)$$

$$\Gamma = \frac{2\gamma}{\gamma-1} - 1$$

Hence,

$$\Gamma = \frac{\gamma+1}{\gamma-1}. \tag{2.42}$$

This is the first important consequence of strong shock which relates the density ratio to the specific heat ratio. Also, from the equation of state, we have

$$\frac{P_2}{\rho_2} = \frac{R_o T_2 z}{M}$$

where $z = 1 + \alpha_1 + 2\alpha_2 + \cdots$ is the departure coefficient.

Combining with Eqs. (2.41) and (2.42), we can obtain

$$T_2 = \left(\frac{M}{R_o}\right)\frac{2(\gamma-1)}{(\gamma+1)^2 z} q_1^2 \tag{2.43}$$

This relates the temperature of the shock heated plasma to the shock speed. Write the kinetic energy per unit mass of the shock heated plasma as

$$k_2 = \frac{1}{2}v_p^2 = \frac{1}{2}(q_1 - q_2)^2 = \frac{1}{2}q_1^2\left(1 - \frac{1}{\Gamma}\right)^2$$

and the enthalpy per unit mass from Eq. (2.39),

$$h_2 = \frac{1}{2}\left(q_1^2 - q_2^2\right) = \frac{1}{2}q_1^2\left(1 - \frac{1}{\Gamma^2}\right)$$

Hence the ratio of enthalpy over kinetic energy

$$\frac{h_2}{k_2} = \frac{1 - \frac{1}{\Gamma^2}}{\left(1 - \frac{1}{\Gamma}\right)^2} = \frac{\Gamma+1}{\Gamma-1}$$

Substitute Γ from (2.42) gives

$$\frac{h_2}{k_2} = \gamma \tag{2.44}$$

This means that in a strong shock heated plasma, the enthalpy is given by specific heat ratio times the kinetic energy.

We consider the example of a strong shock heated hydrogen plasma such that fully ionized state has been achieved. For this plasma, we have $\gamma = \frac{5}{3}$, $\chi = 0$, $\alpha = 1$, hence $z = 1 + \chi + 3\alpha = 4$.

For this fully ionized hydrogen plasma, we can assume that the enthalpy is approximately equal to the ionization potential, which is equal to 15.9 eV (only 1 electron to be ionized per H-atom). While for each H-atom, its kinetic energy when being pushed by the piston to a speed of v_p is

$$k = \frac{1}{2} \times 1.67 \times 10^{-27} v_p^2 \approx 5.3 \times 10^{-9} v_p^2 \text{ (in unit of eV)}$$

From (2.44), $15.9 = \frac{5}{3}\left(5.3 \times 10^{-9} v_p^2\right)$, hence $v_p = 4.3 \times 10^4$ m/s

This means that the hydrogen gas must be driven to a speed of more than 4.3×10^4 m/s for it to be fully ionized.

The shock front speed corresponding to this piston can be obtained from

$$v_{sf} = q_1 = q_2 + v_p = \frac{q_1}{\Gamma} + v_p$$

Hence

$$q_1 = \frac{v_p}{\left(1 - \frac{1}{\Gamma}\right)} = \frac{\Gamma}{\Gamma - 1} v_p = \left(\frac{\gamma + 1}{2}\right) v_p$$

For $\gamma = \frac{5}{3}, q_1 = v_{sf} \approx \frac{4}{3} v_p = 5.7 \times 10^4 \text{ m/s}$.

The temperature of the plasma can also be calculated from (2.43) as

$$T_2 = \left(\frac{M}{R_o}\right) \frac{2(\gamma - 1) q_1^2}{(\gamma + 1)^2 z} = \frac{3}{64} \left(\frac{M}{R_o}\right) q_1^2$$

2.4.2.2 Pulsed Plasma Systems with Shock Heating as Plasma Heating Mechanism

There are numerous types of pulsed plasma systems in which shock heating is operational. These include the electromagnetic shock tube, the Z-pinch and the θ-pinch.

(a) The electromagnetic shock tube.
 In the electromagnetic shock tube, the discharge is initiated at the back-wall uniformly across a cylindrical insulator between the coaxial electrodes to form a current sheet as shown in Fig. 2.29.

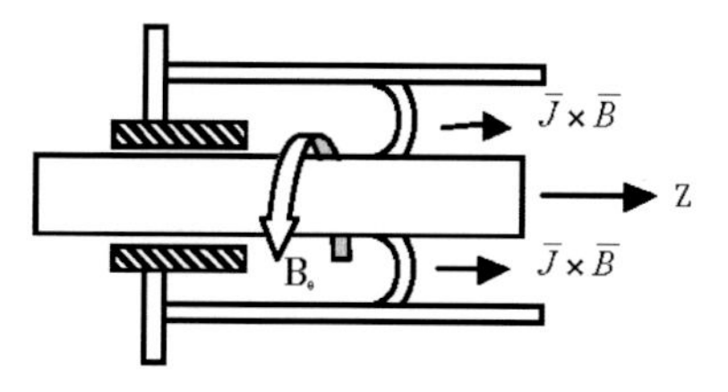

Fig. 2.29 Schematic of an electromagnetic shock tube

The current sheet acts as the electromagnetic piston which is formed first along the surface of the insulator. It will be pushed outwards from the surface of the insulator to the inner surface of the outer electrode by its own electromagnetic force, or the $\boldsymbol{J} \wedge \boldsymbol{B}$ force. The azimuthal magnetic field B_θ is generated by the coaxial discharge current itself and is given by

$$B_\theta = \frac{\mu I}{2\pi r},$$

which is a function of time and radial position r. It can be seen that its magnitude will be stronger at the surface of the inner electrode as compared to the inner surface of the outer electrode. This gives rise to the slanting structure of the electromagnetic piston as shown in Fig. 2.29. The $\boldsymbol{J} \wedge \boldsymbol{B}$ force is in the direction downstream of the tube (in the z-direction) and its magnitude can be expressed as

$$\int_a^b \frac{B_\theta^2}{2\mu} 2\pi r dr,$$

where a is the radius of the inner electrode and b is the radius of the outer electrode. This force will drive the electromagnetic piston to supersonic speed so that a shock heated layer of plasma will be formed. In this way, with discharge current in the region of 100 kA, piston speed of up to more than 10×10^4 m/s may be achieved, which is sufficient to produce fully ionized hydrogen plasma.

(b) The Z-pinch and θ-pinch
In the linear Z-pinch, the electromagnetic piston is cylindrical while the $\boldsymbol{J} \wedge \boldsymbol{B}$ force is in the inward radial direction as illustrated in Fig. 2.30.

Here again $B_\theta = \frac{\mu I}{2\pi r}$ with r is now time varying and the $\boldsymbol{J} \wedge \boldsymbol{B}$ force is

$$F_m = \frac{B_\theta^2}{2\mu}(2\pi r\ell).$$

A variation of the pinch configuration is the θ-pinch which in contrast to the Z-pinch, the discharge current is induced in the θ-direction while its self-magnetic

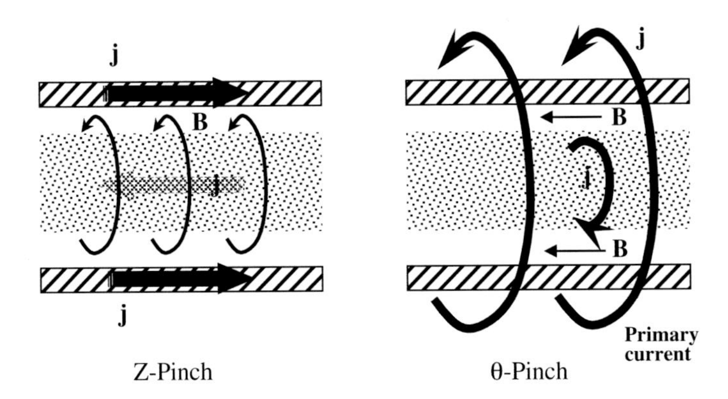

Fig. 2.30 Formation of the Z-pinch and the θ-Pinch

field is in the z-direction instead. The combined effect is also a $\boldsymbol{J} \wedge \boldsymbol{B}$ force in the radially inward direction. This is illustrated in Fig. 2.30, with the Z-pinch configuration shown together for comparison.

References

1. Weston GF (1968) Cold cathode glow discharge tubes (Chap. 1). ILIFFE Books Ltd, London
2. Yip Cheong K (1997) Studies on a planar coil inductively coupled plasma system and its applications. MSc Thesis, University of Malaya
3. Chin OH, Wong CS (1989) A simple monochromatic spark discharge light source. Rev Sci Instrum 60:3818–3819

General References

4. Grey MC (1965) Fundamentals of electrical discharges in gases, part 1, vol 2. In: Beck AH (ed) Handbook of vacuum physics. Bergamon Press, Oxford
5. Reece RJ (1995) Industrial plasma engineering. Principles, vol 1. IOP Publishing Ltd, London

Chapter 3
Plasma Diagnostic Techniques

Abstract In this chapter, some of the basic diagnostic techniques that are useful for studying various types of plasmas are described. These include electrical measurements (discharge current and voltage across the plasma), spectroscopic measurements, the Langmuir (electric) probe, X-ray and neutron measurements.

Keywords Plasma diagnostics

3.1 Electrical Measurements

In many plasma devices such as the plasma focus and the vacuum spark, plasma heating is achieved by the passage of intense current pulse through the plasma. The heating mechanism involved may be the magnetic compression (which is inductive in nature) and/or the joule-heating (which is resistive). In any case, we may consider the plasma as an active element in the discharge circuit with its electrical properties represented by the combination of a variable resistor and a variable inductor. This concept is illustrated in Fig. 3.1.

In this circuit, the energy is initially stored in the capacitor *C*. Upon closing the switch, the capacitor discharges through the circuit. The discharge current is significantly affected by the condition of the plasma that a measurement of the discharge current will be able to reveal information concerning the dynamic changes of the plasma condition during the discharge. Similarly, the transient voltage across the plasma is also directly related to the plasma condition. Combined interpretation of the measured current and voltage waveforms is usually sufficient for basic dynamics study of a pulsed plasma device.

C.S. Wong and R. Mongkolnavin, *Elements of Plasma Technology*,
SpringerBriefs in Applied Sciences and Technology,
DOI 10.1007/978-981-10-0117-8_3

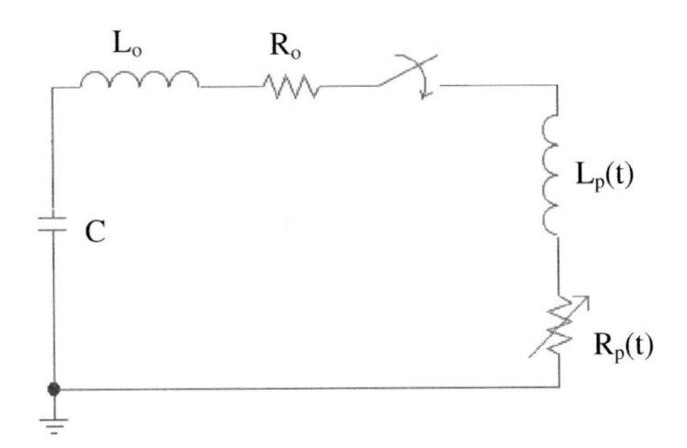

Fig. 3.1 Schematics of a pulsed plasma discharge circuit powered by capacitor discharge

3.1.1 Pulsed Current Measurement by Using the Rogowski Coil

A Rogowski coil is essentially a multi-turn solenoid which is bent into the shape of a torus encircling the current as shown in Fig. 3.2.

In order to understand the principle of operation of the Rogowski coil, we assume the ideal situation where the current to be measured $I(t)$ is passing through the centre of the major cross-section of the torus. Then by Ampere's Law, the magnetic field induced by the current at the axis of the minor cross section of the torus is given by

$$B(t) = \frac{\mu_o I}{2\pi a} \tag{3.1}$$

Then the magnetic flux that threads the minor cross-section of the torus is

$$\phi(t) = \left(\frac{\mu_o A}{2\pi a}\right) I(t) \tag{3.2}$$

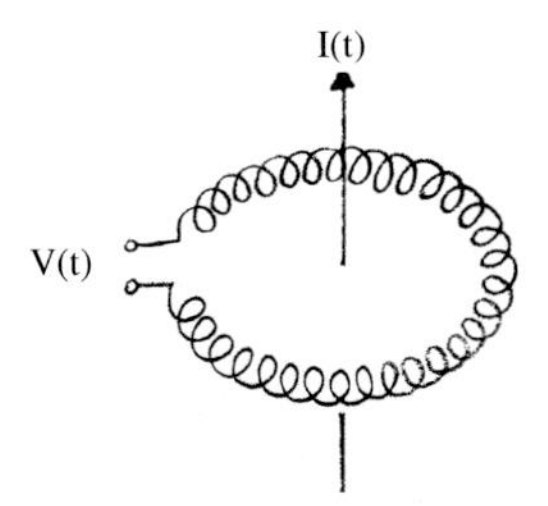

Fig. 3.2 Principle of Rogowski coil for pulsed discharge current measurement. *N* Number of turns, *A* Minor cross-sectional area, *a* Major radius of coil

The voltage induced across the terminal of the coil is

$$V(t) = \left(\frac{\mu_o AN}{2\pi a}\right)\frac{dI}{dt} \tag{3.3}$$

We can see that the induced voltage is proportional to the rate of change of the current, not the current itself. In order to obtain the *I*(*t*), the coil output voltage must be integrated. This can be done in two ways.

3.1.1.1 Integration Using Passive Integrator

The coil output voltage can be integrated by a simple passive *RC* integrator as illustrated in the equivalent circuit shown in Fig. 3.3.

The Rogowski coil is represented by an inductance *L*. The circuit equation can be written as

$$V(t) = \frac{d\phi}{dt} = L\frac{di}{dt} + iR + \frac{1}{C}\int_0^t idt \tag{3.4}$$

where i is the induced current flowing in the coil circuit. We assume that $R \gg L\omega$, then

$$V(t) = iR + \frac{1}{C}\int_0^t idt. \tag{3.5}$$

Further assume that *RC* is chosen such that it is much larger than the characteristic time of the expected plasma event, then we obtain

$$V(t) \approx iR \quad \Rightarrow \quad i \approx \frac{V(t)}{R} = \left(\frac{\mu_o AN}{2\pi a}\right)\frac{dI}{dt}\frac{1}{R}.$$

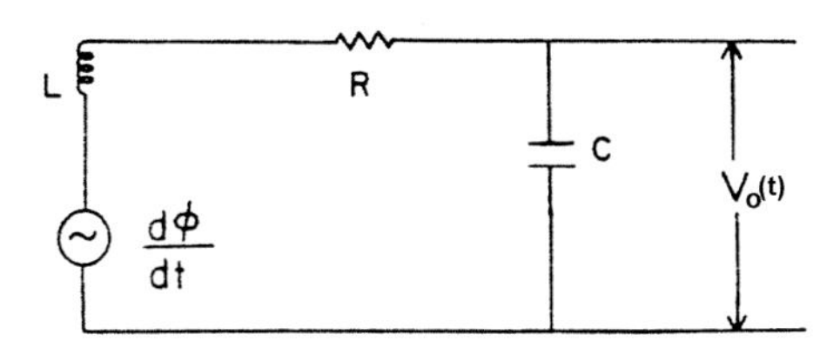

Fig. 3.3 Schematic circuit diagram of Rogowsdi coil with RC integrator

The integrated output of the coil is measured across C and hence given by

$$V_o(t) = \frac{1}{C}\int_0^t idt = \frac{\mu_o NA}{2\pi aR} I(t) \propto I(t) \tag{3.6}$$

which is directly proportional to the discharge current itself.

For this mode of operation, the Rogowski coil has a lower frequency limit of

$$f > \frac{1}{RC} \tag{3.7}$$

while its upper limit is set by the fact that a 50 Ω coaxial cable is normally required to connect the coil to the RC integrator which is in turn connected directly to the oscilloscope input. In this case the coaxial cable is shorted to ground via a 50 Ω resistor before connected to the RC integrator. It is required that $\omega L < 50$. This gives rise to the upper frequency limit for the Rogowski coil as

$$f < \frac{50}{2\pi L} \quad \text{or simply} \quad f < \frac{50}{L}. \tag{3.8}$$

This limit on the frequency response of the Rogowski coil can be quite serious since

$$\text{if } L \sim \mu\text{H}, \quad \Rightarrow \quad f < 50 \text{ MHz}.$$

3.1.1.2 The Rogowski Coil as a Current Transformer

The Rogowski coil can be operated as a current transformer by shorting its terminals by a small resistance R as shown in Fig. 3.4.

The circuit equation in this case is

$$V(t) = L\frac{di}{dt} + i(r + R) \tag{3.9}$$

where r is the resistance of the coil which is normally negligible and L is the coil inductance. If R is chosen to be small such that $(r + R) \ll \omega L$, then

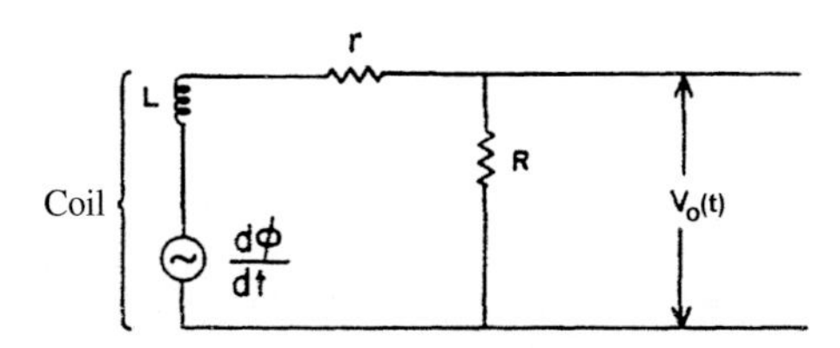

Fig. 3.4 Schematic circuit diagram of Rogowski coil operated as current transformer

$$V(t) = L\frac{di}{dt} = \frac{d\phi}{dt} \tag{3.10}$$

and hence

$$i(t) = \frac{1}{L}\int_0^t V(t)dt = \frac{\phi}{L} \tag{3.11}$$

with

$$\phi = \frac{\mu_o NAI(t)}{2\pi aR}$$

and

$$L = \frac{\mu_o N^2 A}{2\pi aR}$$

$$V_o(t) = iR = \frac{R}{N}I(t) \tag{3.12}$$

which is the output voltage measured across R.

In principle, the frequency response of the Rogowski coil operated as current transformer is only limited by the transit time of electrical signal along the length of the coil which is given by $2\pi a$. The transit time along the coil is given by

$$\tau = \sqrt{lc} \cdot 2\pi a \tag{3.13}$$

where l is the inductance per unit length while c is the capacitance per unit length of the coil. This gives the upper frequency limit of the Rogowski coil operated as current transformer as

$$f < \frac{1}{\tau}. \tag{3.14}$$

The lower frequency limit is set by the condition that $R < \omega L$ which gives

$$f > \frac{R}{L}. \tag{3.15}$$

3.1.1.3 Calibration of the Rogowski Coil

For absolute measurement of the discharge current, the Rogowski coil must be calibrated. The most straight forward method of calibrating the coil is to use Eq. (3.6) or (3.12). However, it is difficult to estimate the geometry of the coil (N, A, and a) or the value of the small resistance used to short the coil (R) accurately. Hence this method of calculating the calibration factor is seldom practiced. The Rogowski coil calibration is usually done by the so call in situ method. In this case, an ideal *LCR* discharge circuit with light damping is used. The method is said to be "in situ" since the coil can be calibrated at the position it is being used for the actual experiment. An experimental condition will be chosen such that an ideal *LCR* discharge is obtained. An example of the discharge current waveform for such a discharge is as shown in Fig. 3.5.

This current waveform can be represented mathematically by:

$$I(t) = I_o e^{-\alpha t} \sin \omega t \tag{3.16}$$

where

$$I_o = V_o \sqrt{\frac{C}{L}}, \quad \alpha = \frac{R}{2L}, \quad \omega = \frac{1}{\sqrt{LC}}.$$

From the current waveform, we can measure T, V_1, and V_2. Then

$$\alpha = -\frac{\ell n(V_2/V_1)}{T} \tag{3.17}$$

Consider the first peak current at $t = \frac{T}{4}$

$$I_1 = I_o e^{-\alpha T/4} \tag{3.18}$$

which can be calculated since I_o, α and T can be obtained. Hence the calibration factor of the coil can be obtained as

$$K = \frac{I_1}{V_1} = \frac{2\pi \, CV_o}{TV_1} \left(\frac{V_2}{V_1}\right)^{\frac{1}{4}}. \tag{3.19}$$

In most pulsed plasma experiments, the discharge current generated is an under damped sinusoidal waveform. The main plasma heating mechanism normally

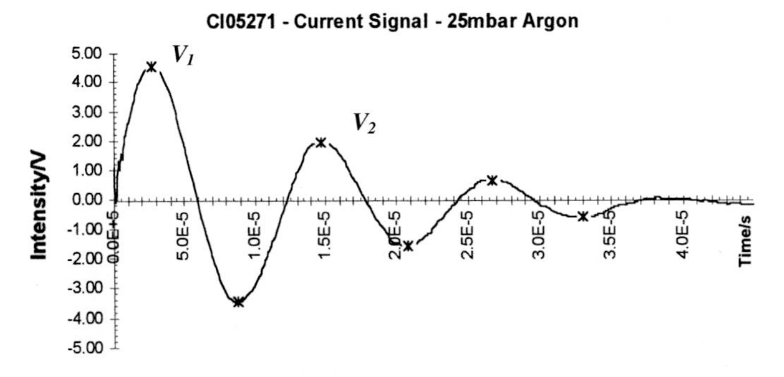

Fig. 3.5 Under-damped sinusoidal discharge current waveform

occurs during the first half cycle of the discharge where the discharge current waveform is severely deformed from that of an ideal sinusoidal waveform. For example, in a plasma focus discharge, the plasma heating which occurs first through the axial acceleration phase and then the radial compression phase during the first half cycle of the discharge. This implies that the plasma impedance may be changing and hence the resulting discharge current waveform may deviate from that of an ideal *LCR* discharge. For the purpose of Rogowski calibration, the plasma focus can be shorted at its back wall by a high power resistor acting as the load. This high power resistor can be constructed by using, for instant, copper sulphate solution. Alternatively, the plasma focus can be discharged under high ambient pressure condition so that the current sheath will hardly move during the first few cycles of the discharge current. The discharge current waveform can be acquired at a compressed time scale so that several cycles of the waveform can be recorded as shown in Fig. 3.5.

3.1.2 Pulsed Voltage Measurements

It is informative if the transient voltage across the plasma can be measured. This can be done by using either the resistive voltage divider or the capacitive voltage divider.

3.1.2.1 The Resistive Voltage Divider

The principle of resistive voltage divider is very simple, as illustrated in Fig. 3.6.

The high input voltage V_i is divided by R_1 and R_2 and the output voltage is measured across the lower resistance R_2 which is given by

$$V_o = \frac{R_2}{R_1 + R_2} V_i.$$

A practical design of the resistive voltage divider is shown in Fig. 3.7.

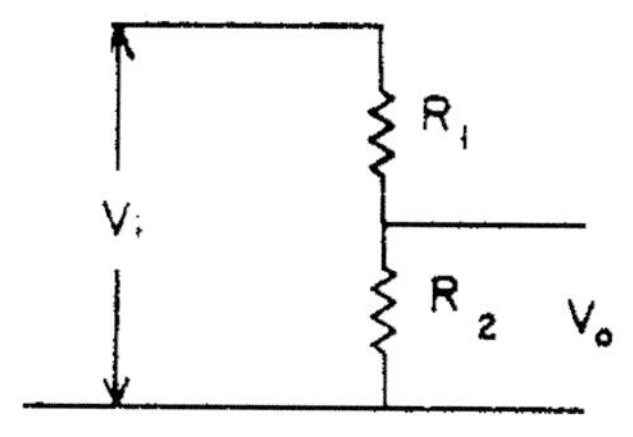

Fig. 3.6 Schematics of a resistive voltage divider

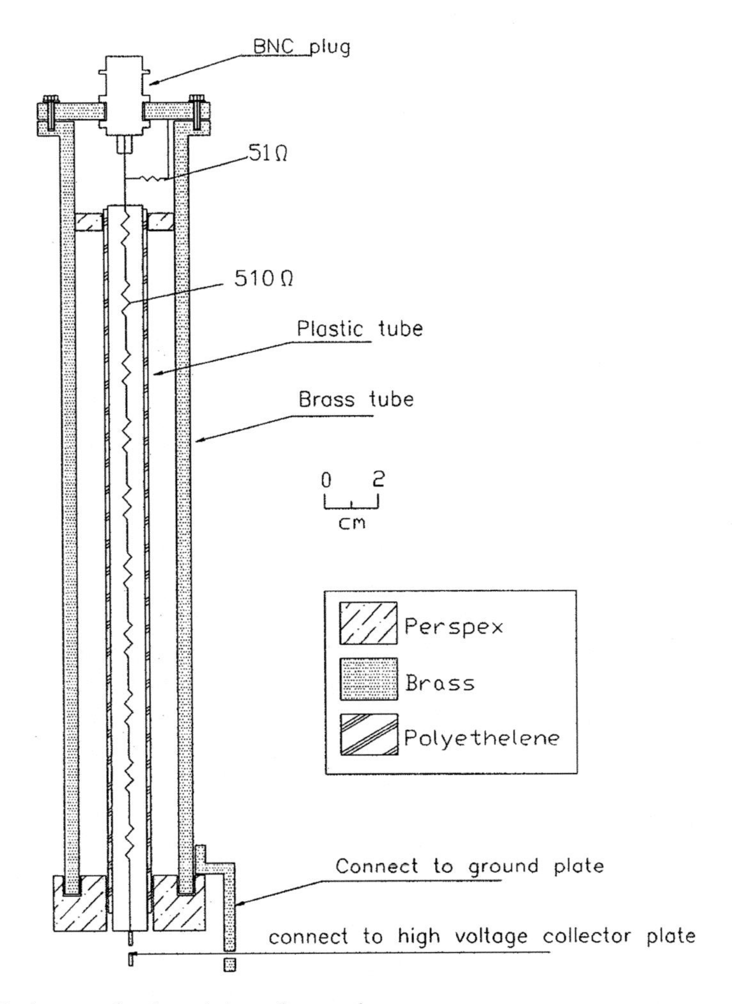

Fig. 3.7 An example of a resistive voltage probe

While the principle of the resistive divider is rather simple, there are several technical difficulties in the implementation of the divider to measure the voltage across the plasma: (i) It may be difficult to connect the divider directly across the plasma so the voltage measured may not be the true voltage across the plasma but consists of additional voltage contributed by other stray components of the discharge circuit. (ii) In most cases the voltage divider is connected directly onto the high voltage point to be measured so it is necessary to use long coaxial cable to transmit the signal to the oscilloscope which may be placed inside a screened room or box. This restricts the value of R_2 to be 50 Ω which is the impedance of the coaxial cable and the coaxial cable coupled to the 50 Ω terminator which will act as R_2 as a whole. One possible configuration is to choose R_2 to be much lower than 50 Ω, say 1 Ω, but then the attenuation factor will be high since it is required that

$R_1 + R_2$ must be sufficiently large so that they will not act to short the plasma load. For low R_2, the upper frequency limit of the divider may also be affected since the upper frequency limit is determined by

$$f < \frac{R_2}{L_s}$$

where L_s is the stray inductance of the probe circuit which is difficult to avoid.

3.1.2.2 Capacitive Voltage Divider

In the capacitive voltage divider, the voltage division is achieved by using capacitors instead of resistors as shown in Fig. 3.8.

Here the output voltage is measured across C_2 which is of larger value than C_1. This is given by

$$V_o = \frac{C_1}{C_1 + C_2} V_i.$$

It is however technically difficult to implement the capacitive divider since in order to measure the voltage across C_2 accurately the measuring device must have infinite impedance. This is not possible since it is normally necessary to use coaxial cable to connect between the divider and the oscilloscope. This means shorting C_2 with 50 Ω and this will defeat the functioning of the capacitive voltage divider.

A possible configuration to implement the capacitive divider is to connect C_2 directly to the input of the oscilloscope which has an input impedance of 1 MΩ. In this case C_2 is formed by using a coaxial cable which has a capacitance of 100 pF/m, while C_1 is formed between the central conductor of the coaxial cable and the high voltage point to be measured, separated by insulator. This is shown schematically in Fig. 3.9 [1]. For this configuration, one important trick is the oscilloscope input must be set to DC mode to bypass the input capacitor.

The practical usage of this capacitive divider configuration is, however, quite limited. It is probably only suitable for low power (current of a few kA and voltage of a few kV) application. For high pulsed power application, a hybrid

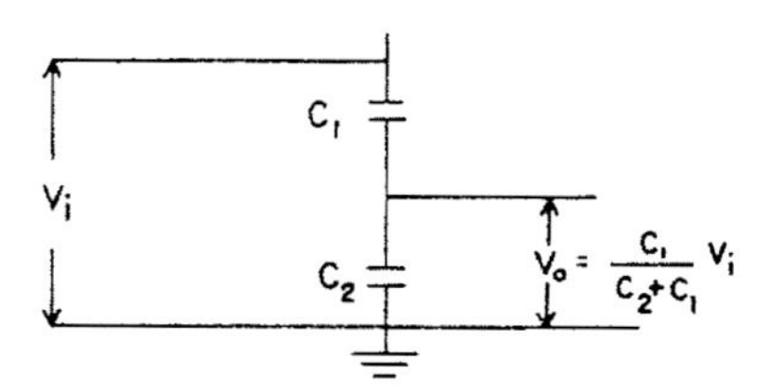

Fig. 3.8 Schematics showing voltage division by capacitors

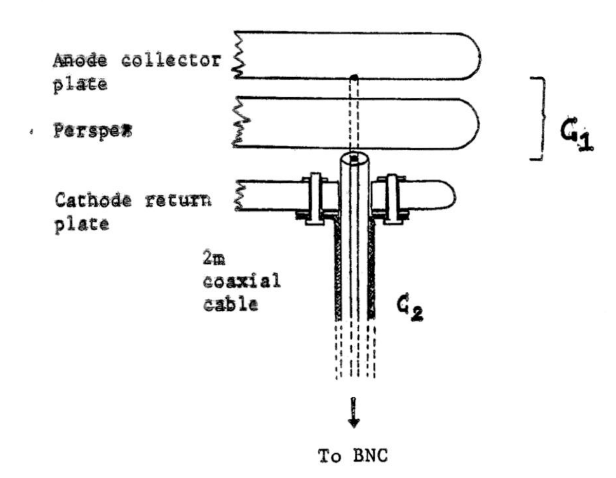

Fig. 3.9 Illustration of the mounting of a simple capacitive voltage divider

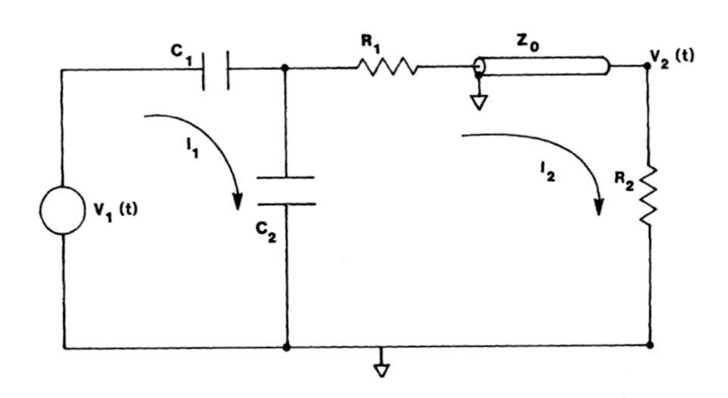

Fig. 3.10 Hybrid capacitive-resistive voltage probe

capacitive-resistive configuration is usually employed. An example of such a configuration is shown schematically in Fig. 3.10.

The overall voltage division of this high voltage probe is given by:

$$\frac{V_o}{V_i} = \frac{R_2}{R_1 + R_2} \frac{C_1}{C_1 + C_2}.$$

It is important to choose the values of R_1 and R_2 so that

$$(R_1 + R_2) \gg \frac{1}{\omega C_2}.$$

This set the frequency limit for the probe to operate in the voltage divider mode (where $V_o \propto V_i$) at

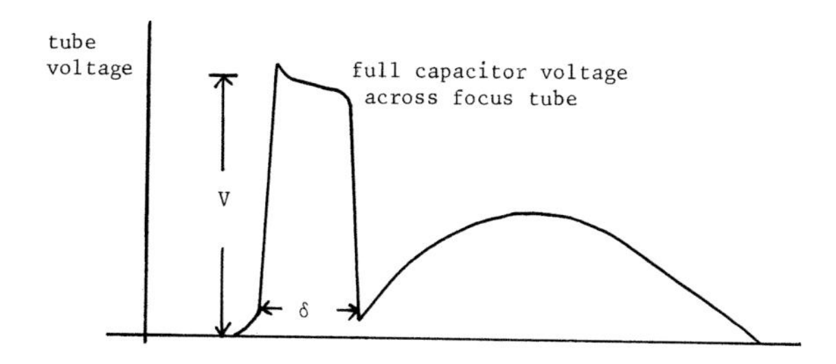

Fig. 3.11 An example of the voltage signal recorded for the discharge of electromagnetic shock tube

$$f > \frac{1}{(R_1 + R_2)C_2}.$$

For frequency below this limit, the probe is expected to operate in the differential mode where

$$V_o \propto \frac{dV_i}{dt}.$$

For frequency $f \approx \frac{1}{(R_1 + R_2)C_2}$, we will get a mixture of the divider and differential modes which will be difficult to interpret.

3.1.2.3 Calibration of the Voltage Divider

Although it is normally possible to estimate the voltage division of the resistive or capacitive divider from the values of the components used, it is useful to perform a calibration of the probe for more accurate measurement. This can be done by using a standard pulse generator with know amplitude and rise and fall times as the voltage source. In situ calibration can also be performed depending on the actual experimental setup. For example, in the case of the electromagnetic shock tube, a suitably delayed breakdown by lowering the filling pressure will give a signal as shown in Fig. 3.11.

Calibration of the voltage divider can then be obtained by dividing the discharge voltage by V measured from this waveform.

3.1.3 Interpretation of the Current and Voltage Waveforms

From the current and voltage waveforms measured, several system parameters of the electrical discharge circuit as well as information concerning the dynamics of the plasma can be deduced.

Before beginning a series of experiments, it is necessary to first determine the basic parameters of the discharge system such as the circuit inductance and resistance. This is done by creating a discharge condition whereby the plasma inductance and resistance are negligible. Referring to the electromagnetic shock tube again, this means shorting the tube at the back so that it is cut off from the discharge circuit. Whatever inductance and resistance that are present now will be due to the circuit external to the tube only. Hence from the discharge current waveform obtained, which should be that of an ideal *LCR* discharge with fixed inductance and resistance, L_o can be estimated from T since $T = 2\pi\sqrt{L_o C}$. Similarly from the measurement of the damping factor α, the stray resistance of the circuit R_o can be estimated since $\alpha = \frac{R_o}{2L_o}$.

With the plasma discharge, if we assume the plasma to be inductive and the plasma resistance is negligibly small such that $R_p \ll \omega L_p$, then

$$V_p(t) = \frac{d}{dt}(L_p I).$$

With $I(t)$ and $V_p(t)$ obtained, $L_p(t)$ can be deduced from

$$L_p(t) = \frac{\int V_p(t)dt}{I(t)}.$$

$L_p(t)$ is related to the dynamics of the plasma. For example in the electromagnetic shock tube,

$$L_p(t) = \frac{\mu_o}{2\pi}\ell n\left(\frac{b}{a}\right)z(t)$$

so the trajectory of the electromagnetic piston (and hence the shock front) can be deduced. Similarly, for the Z-pinch discharge, the radial trajectory of the current sheet can be deduced from

$$r(t) = r_o \exp\left(-\frac{2\pi L_p(t)}{\mu_o \ell}\right).$$

For a resistive plasma such as the high pressure flash lamp where L_p is fixed, $R_p(t)$ can be deduced from

$$R_p(t) = \frac{\left(V_p(t) - L_p\frac{dI(t)}{dt}\right)}{I(t)}.$$

Assuming the plasma to be a uniform column with length l and cross-sectional area A, the resistivity of the plasma can be obtained:

$$\eta_p(t) = R_p(t)\frac{l}{A} = 65.3T_e^{-5/2}(\ell n\Lambda)Z.$$

With an estimate of Z (the charge state of the plasma) and taking $\ell n\Lambda \approx 10$, the temperature of the plasma $T_e(t)$ can be estimated.

3.2 Pulsed Magnetic Field Measurement

In a pulsed electric discharge, since the current is rapidly changing with time, a pulsed magnetic field will be induced. The Rogowski coil is in fact a means to measure this induced magnetic field in order to deduce the discharge current $I(t)$. Since the Rogowski coil is in the form of a solenoid bent into a torus encircling the current, it measures the total discharge current and does not provide information concerning the distribution of the current. In order to obtain information concerning the distribution, or of greater interest the location of the current path such as the location of the current sheet in the electromagnetic shock tube, a magnetic pick-up coil can be used to measure the localised magnetic field.

The magnetic pick-up coil is a small diameter coil of several turns made from insulated copper wire as shown in Fig. 3.12.

The principle of the magnetic pick-up coil is the same as the Rogowski coil, with the output voltage induced given by

$$V(t) = NA\frac{dB(t)}{dt}$$

where N is the number of turns of the coil and A is the cross-sectional area of the coil. $B(t)$ is the time-varying magnetic field induced by the discharge current at the location of the probe.

In order to obtain $B(t)$ from the output voltage waveform of the coil, it is necessary to perform integration as in the case of the Rogowski coil. However, in this case it will not be possible to operate in the current transformer mode since the

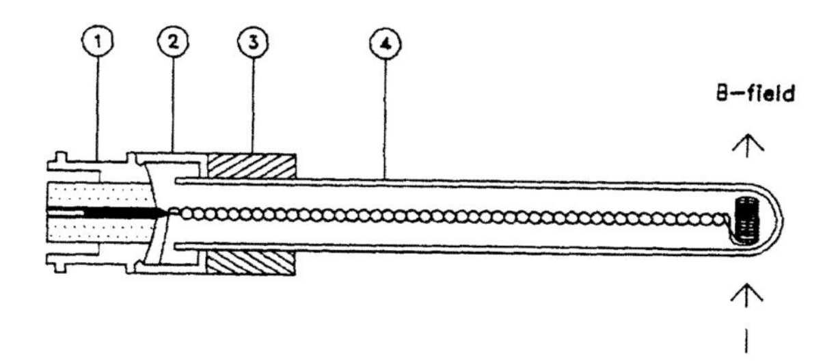

Fig. 3.12 Example of a design of the magnetic pick-up coil. *1* BNC socket. *2* Copper adapter. *3* Perspex holder. *4* Glass tube

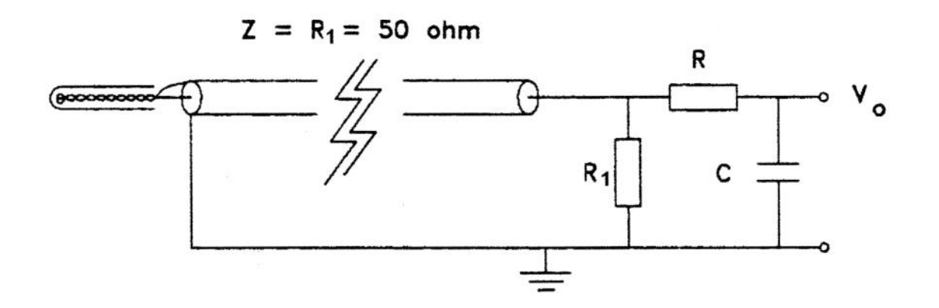

Fig. 3.13 Schematic circuit diagram of a magnetic probe

inductance of the coil is very low so the condition $R \ll \omega L$ cannot be satisfied. As such the magnetic pick-up coil is usually operated with an RC integrator as shown in Fig. 3.13.

In this case the final output voltage obtained will be $V_o(t) = \frac{NA}{RC} B(t)$. The frequency range for the operation of the coil is $\frac{1}{RC} < f < \frac{Z}{L}$.

3.3 Plasma Spectroscopy

In this section, we will discuss briefly on some fundamental concepts and techniques on the measurements of optical radiation emission from plasmas.

3.3.1 Plasma Radiation

The particles (electrons, ions and neutrals) are moving inside the plasma with kinetic energy. Interactions between these particles, most likely involving collisions of electrons with the heavier particles will result in various processes. Some of these processes may lead to emission of photons in a wide range of spectrum from infrared up to gamma ray depending on the temperature of the plasma. There are basically three types of processes that will give rise to the emission of radiation from the plasma, namely Bremsstrahlung, recombination and radiative decay.

3.3.1.1 Bremsstrahlung

In a plasma, electrons are moving under the effect of the electromagnetic field of the bulk of the particles inside the plasma. They may encounter interaction that causes them to be retarded. This retardation results in the release of one quantum of energy (one photon) by the electron. This process is called Bremsstrahlung. Since the electron emitting the photon is in the free state before and after the interaction, the transition is often called a "free-free transition". Bremsstrahlung may give rise to photon of continuously varying energy and hence contributed to the continuum of the photon energy spectrum.

The intensity of Bremsstrahlung emission (J per m^3 per second) in the frequency interval of v to $v + dv$ caused by the interaction of N_e electrons/m^3 with N_i ions/m^3 is given by

$$\frac{dE_{ff}}{dv} = CN_eN_iZ_i^2\left(\frac{\chi_H}{kT_e}\right)^{1/2}\bar{g}_{ff}\exp\left(-\frac{hv}{kT_e}\right) \quad (\mathrm{W\,m^{-3}}) \tag{3.20}$$

where $C = 1.7 \times 10^{-53}$ J m^3, Z_i is the charge number of the ions, χ_H is the ionization potential of hydrogen, and $\bar{g}_{ff}$ is the free-free Gaunt factor which represents the departure of the quantum-mechanical calculation from the classical result averaged over the Maxwell-Boltzmann velocity distribution at electron temperature T_e. It is normally more convenient to express the emission in terms of wavelength, and Eq. (3.20) will then becomes

$$\frac{dE_{ff}}{d\lambda} = CN_eN_iZ_i^2\left(\frac{\chi_H}{kT_e}\right)^{1/2}\bar{g}_{ff}\frac{c}{\lambda^2}\exp\left(-\frac{hc}{\lambda kT_e}\right). \quad (\mathrm{W\,m^{-3}\,\AA^{-1}}) \tag{3.20a}$$

If we consider that ionic species with charge states of Z_m, Z_{m+1}, … Z_n are present inside the plasma and their fractional abundances are α_m, $\alpha_{m+1}, \ldots \alpha_n$ then the average charge state of the plasma can be taken as

$$Z_{eff} = \frac{\sum_{j=m}^{n}\alpha_j Z_j}{N_T} \tag{3.21}$$

The total Bremsstrahlung intensity can be calculated by using Eq. (3.20a) with $Z_i = Z_{eff}$. α_j's can be calculated using suitable plasma model such as LTE (Local Thermodynamic Equilibrium) or CE (Coronal Equilibrium).

3.3.1.2 Recombination

Recombination is a process when a free electron of kinetic energy ε_e collides with an ion (atomic species s and charge Z_i) is captured in a bound level (principal quantum number n and ionization potential χ_{i-1}^n) of the ion of charge ($Z_i - 1$) resulting in the emission of a photon of energy $hv = \varepsilon_e + \chi_{i-1}^n$. It can be seen that the photon energy is a function of the electron energy which is continuous, and only photons with energy greater than χ_{i-1}^n will be emitted. This results in discontinuities, called recombination edges in the free-bound continuum spectrum.

The intensity of recombination emission (J per m^3 per second) in the frequency interval of v to $v + dv$ caused by the interaction of N_e electrons/m^3 with N_i ions/m^3 (assuming recombination into the nth shell of a hydrogen like ion of charge i) is given by

$$\frac{dE_{fb}}{d\nu} = CN_eN_i\left(\frac{\chi_H}{kT_e}\right)^{3/2}\left(\frac{\chi_{i-1}^n}{\chi_H}\right)^2\frac{\zeta_n}{n}\bar{g}_{fb}\exp\left(-\frac{\chi_{i-1}^n - h\nu}{kT_e}\right) \quad (\mathrm{W\,m^{-3}}) \tag{3.22}$$

where χ_H is the ionization potential of hydrogen atom; χ_{i-1}^n is the ionization potential of ionic specie at charge state of Z_{i-1} from bound state with principal quantum number n; and ζ_n is the number of vacancies available for recombination with electron at that bound state. C is the same constant as in expression (3.20). Expressed in terms of wavelength, it becomes

$$\frac{dE_{fb}}{d\lambda} = CN_eN_i\left(\frac{\chi_H}{kT_e}\right)^{3/2}\left(\frac{\chi_{i-1}^n}{\chi_H}\right)^2\frac{\zeta_n}{n}\bar{g}_{fb}\frac{c}{\lambda^2}\exp\left(-\frac{\chi_{i-1}^n - \frac{hc}{\lambda}}{kT_e}\right). \quad (\mathrm{W\,m^{-3}\,\mathring{A}^{-1}}) \tag{3.22a}$$

Similarly, the total intensity of the recombination emission can be obtained by consideration of the contribution from various ionic species present.

3.3.1.3 Radiative Decay

An atom or ion collided by electron may be excited and subsequently undergoes radiative decay which results in the emission of a photon. The energy of this photon corresponds to the energy difference between the two energy levels so it is characteristic of the emitting atom or ion and it has a discrete value. The transition is between two bound states and it is called bound-bound transition.

The intensity of line emission (J per m^3 per second) at the frequency ν due to transition from level p to level q is given by

$$I_{pq} = (h\nu)A(p,q)N(p) = \frac{hc}{\lambda}A(p,q)N(p) \quad (\mathrm{J\,m^{-3}}) \tag{3.23}$$

where $A(p, q)$ is the transition probability, $N(p)$ is the number density of the upper energy state.

3.3.1.4 Some Useful Facts About Plasma Emission Spectrum

(i) The shapes of the Bremsstrahlung and recombination emission spectra are identical except at the recombination edges.
(ii) The peak of the plasma emission continuum occurs at wavelength of

$$\lambda_o = \frac{6200}{T_e} \quad (\mathring{A}), \tag{3.24}$$

where T_e is in eV.

(iii) The contribution of Bremsstrahlung towards the continuum will be dominant at high temperature. That is

$$\frac{dE_{ff}}{d\nu} \gg \frac{dE_{fb}}{d\nu} \quad \text{when} \quad kT_e > 3Z_i^2\chi_H. \tag{3.25}$$

3.3.2 *The Plasma Models*

In order to calculate the radiation spectrum emitted from the plasma, it is necessary to know what are the species present and what is their fractional distribution. The real plasma state is extremely complicated that no one model is able to give an accurate and complete description. In this section we will discuss two approximate plasma models.

3.3.2.1 Local Thermodynamic Equilibrium (LTE)

Ideally, life will be easier if a state of complete thermodynamic equilibrium (TE) can be satisfied by the plasma. Then the plasma state can be described by a set of finite number of thermodynamic variables such as the temperature, the pressure and the concentrations of various elements present while the plasma emission can be approximated to the blackbody radiation. One necessary condition of this situation is that the whole plasma is homogeneous and radiation emitted from the interior of the plasma will be completely re-absorbed by the plasma itself. In reality this is not the case. The best that can be achieved is to divide the volume of the plasma into small elements of volume so that the plasma inside each element can be considered to be homogeneous and yet its dimension is larger than the mean free path of the particle or photon. This the case of the LTE plasma model.

We consider the ionization of the ith ionization state of the element when it is collided by electron inside the plasma concerned. If the pressure (and particle density) of the plasma is sufficiently high, three body recombination will occur and balance the ionization process:

$$A_i + e \Leftrightarrow A_{i+1} + e + e$$

Assuming that all species are present at their ground state only, the ratio of the densities of two species with consecutive ionic states (i and i + 1) is given by the Saha equation:

$$\frac{N_{i+1}}{N_i} = \frac{2}{N_e}\left(\frac{U_{i+1}}{U_i}\right)\left(\frac{2\pi m_e kT_e}{h^2}\right)^{3/2} \exp\left(-\frac{\chi_i}{kT_e}\right) \tag{3.26}$$

where U's are the respective partition functions of the two species and E_i is the ionization potential [for ionization from ith ionic state to $(i + 1)$th state]. Within each specie, the distribution of population densities among the bound levels is given by the Boltzmann relation:

$$\frac{N(p)}{N(q)} = \frac{g_p}{g_q}\exp\left(-\frac{E_p - E_q}{kT_e}\right) \tag{3.27}$$

where p and q refer to two energy levels with energies E_p and E_q and statistical weights g_p and g_q respectively. For each level, the number density can be written as:

$$N(p) = N\frac{g_p}{U_e}\exp\left(-\frac{E_p}{kT_e}\right) \tag{3.28}$$

where U_e is the electronic partition function of the specie which is given by

$$U_e = \sum_{all\, j} g_p \exp\left(-\frac{E_j}{kT_e}\right); \tag{3.29}$$

and N is the total number density given by

$$N = \sum_{all\, j} N(j). \tag{3.30}$$

Note that in (3.26), U_i is the electronic partition function of the specie with ith ionic state.

In addition to the three body recombination that occurs as the inverse of electron collisional ionization, radiative recombination is also taking place simultaneously. However, in a high density plasma, the probability for radiative recombination to occur is very much lower than the three body recombination. This condition can be shown to be true when

$$N_e \geq 1.6 \times 10^{12} T_e^{1/2} E_{pq}^3 \text{ cm}^{-3} \tag{3.31}$$

where E_{pq} is the highest energy gap in eV, normally the energy gap between the first two levels. T_e is the electron temperature in K.

3.3.2.2 Corona Equilibrium (CE) Model

In a low density plasma, where collision does not occur with sufficiently high frequency to enable the establishment of thermodynamic equilibrium (not even local), the plasma must be described by non-thermal or non-LTE model. One of the frequently used model which was first proposed to explain certain features of the solar corona is the Coronal Equilibrium Model. It is assumed that the changes in the plasma condition occur sufficiently slowly so that the electrons are always at thermal equilibrium among themselves so that they will have a Maxwellian velocity distribution.

Again assuming that the ionic species are present in their ground states only, and electron collisional ionization is always balanced by radiative recombination, then we may write

$$N_e N_i S(T_e, i) = N_e N_{i+1} R(T_e, i+1) \tag{3.32}$$

where $S(T_e, i)$ is the collisional ionization coefficient and $R(T_e, i + 1)$ is the radiative recombination coefficient. Hence the ratio of the densities of two consecutive ionic species can be obtained from

$$\frac{N_{i+1}}{N_i} = \frac{S(T_e, i)}{R(T_e, i+1)}. \tag{3.33}$$

The expressions for $S(T_e, i)$ and $R(T_e, i + 1)$ are obtained empirically and substituting them into Eq. (3.33) gives rise to

$$\frac{N_{i+1}}{N_i} = 1.27 \times 10^8 \frac{1}{\chi_i^2} \left(\frac{kT_e}{\chi_i} \right)^{3/4} \exp\left(-\frac{\chi_i}{kT_e} \right). \tag{3.34}$$

As we mentioned earlier, it was assumed that the plasma condition changes slowly so that equilibrium can be achieved. The time for the plasma to achieve this equilibrium is the atomic relaxation time given by

$$\tau \approx \frac{10^{12}}{N_e}. \text{ (s)} \tag{3.35}$$

If the plasma lifetime is shorter than this relaxation time, a time-dependent Corona Model may be required to describe the plasma. On the other hand, if it is desired to include effect of electron collision processes causing transition between upper levels, the Collisional Radiative (CR) model may be appropriate.

In spectroscopic experiments of plasma, it is often required to base the analyses of data on a particular plasma model. For simplicity, we may decide to choose between either the LTE model or the CE model. In this case we can base our choice on expression (3.31). This means that if the density of the plasma is high and satisfy the condition stated in (3.31), we choose LTE model. Otherwise, the CE model will be used. However, we should remember that this treatment is not accurate.

3.3.3 *Examples of Population Density Distribution and Plasma Spectra*

3.3.3.1 Population Density Distribution

For the following examples, the fraction of ionization is defined as $\alpha_i = \frac{N_i}{N_t}$ where $N_t = \sum_{all\ i} N_i$, with i = 0 denoting neutral. Figure 3.14a, b show the population density distributions for argon and carbon plasmas according to the CE model.

We can see from the comparison of these two sets of graphs that the argon plasma is fully ionized at electron temperature above 10 keV while the carbon plasma is almost fully ionized at an electron temperature of above 200 eV.

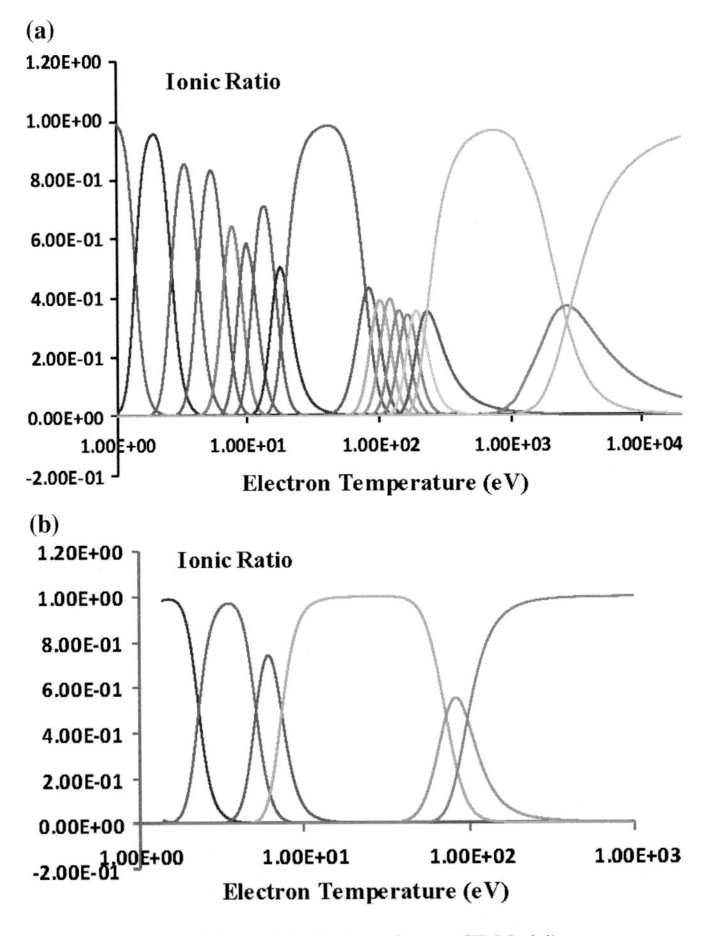

Fig. 3.14 **a** Argon plasma (CE Model). **b** Carbon plasma (CE Model)

3.3.3.2 Examples of Plasma Emission Spectra

Figure 3.15 shows the spectrum of an argon plasma at an electron temperature of 1 keV and electron density of 10^{19} cm^{-3}. All three types of emission: Bremsstrahlung, recombination and line radiation are included. The spectrum is computed assuming CE model which predicts that at this temperature the most prominent species are the Ar^{16+} (Helium-like) and Ar^{17+} (Hydrogen-like) with characteristic lines in the X-ray region. The peak of the continuum is expected to be at about 6 Å. When the temperature is increased to 2 keV, it is clear that the spectrum has shifted to shorter wavelength, while the recombination edge and the line radiation are not affected in terms of their wavelengths (Fig. 3.16).

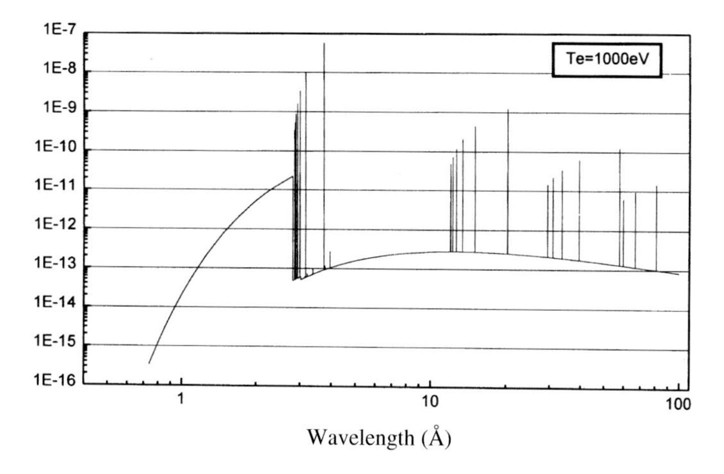

Fig. 3.15 Simulated spectrum of an argon plasma at electron temperature of 1 keV

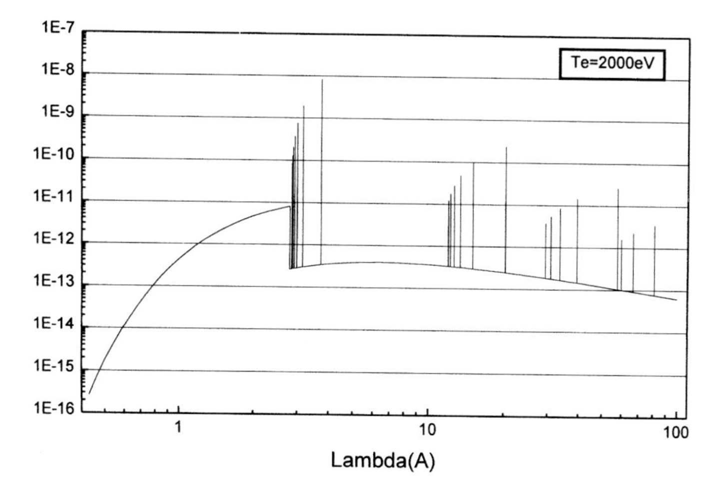

Fig. 3.16 Simulated spectrum of an argon plasma at electron temperature of 2 keV

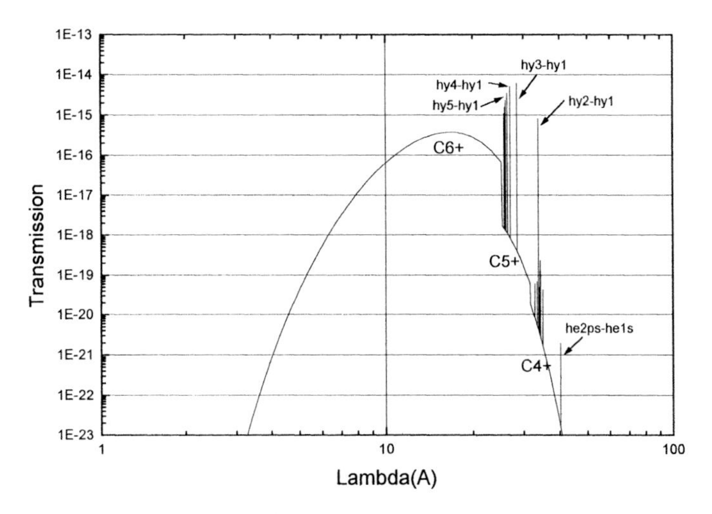

Fig. 3.17 Simulated spectrum of a carbon plasma at electron temperature of 150 eV

Figure 3.17 shows the example of the emission spectrum of carbon plasma at 150 eV computed assuming CE model.

For this plasma, the prominent species are clearly C^{6+} and C^{5+} together with small fraction of C^{4+}.

3.3.4 Optical Emission Spectroscopy of Plasma

In principle, the electron temperature of the plasma can be estimated from the peak of the continuum if it can be determined with reasonable accuracy. This is normally difficult to accomplish particularly for low temperature plasma where the spectrum is recombination and line radiations dominated. One of the most commonly used method to determine the electron temperature of low temperature plasma is the line ratio method. This involves the measurement of line radiation emission spectrum of the plasma (normally in the visible region for low temperature plasma) and the electron temperature can be deduced from the ratio of a pair of nearby spectra lines.

For a continuous plasma such as the DC glow or arc discharge, a simple setup consisting of a monochromator (also known as spectrograph) coupled to a single channel photomultiplier tube will be sufficient. In this case the light emitted from the plasma is collected and fed to the monochromater through an input (or entrance). After entering the monochromator, the light is reflected and focussed onto a grating which acts as the dispersion medium. The dispersed light of various wavelengths will then be collected at the exit window. The range of wavelengths that will appear in the exit window is selected by tilting the plane of the grating at different angle. The overall range of wavelengths that may be detected by the monochromator, as well as its resolution, will be determined by the groove density (grooves/mm) of the grating.

For the setup using single channel photomultiplier tube to measure the spectral intensity as a function of wavelength, an exit slit is used to limit the view of the photomultiplier so that only photons of a particular wavelength will be detected. Since the width of the slit is finite, this will introduce the second factor of instrumental width to the spectral line. The entrance slit gives rise to the first factor of instrumental spectral width. The mechanism for the tilting of the grating is connected to a dial with reading calibrated to give the wavelength directly. When the grating is being tilted, each of its angular position is related to the wavelength of the photon as given by the diffraction formula $2d \sin \theta = n\lambda$ so that at each angular position, only the wavelength that satisfies this condition will be "seen" by the photomultiplier through the exit slit. Hence the spectrum can be scanned by adjusting the dial for the case of DC plasma. The similar procedure can be followed for the RF plasma as well.

Recently, with the availability of photo-diode-array (PDA) and CCD, optical multi-channel analyser (OMA) is often used instead of the single channel PMT to measure the plasma emission spectrum. Figure 3.18 shows an example of such a setup used for measuring the glow discharge emission spectrum. In this setup a range of the spectrum can be measured simultaneously. The range is limited by the width of the exit window of the monochromater, but more often by the length of the PDA or CCD. The commonly available PDA and CCD have 1024 pixels, which mean 1024 points are acquired simultaneously. With smaller pixel size and closely packed pixels, better resolution can be achieved but we will have to compromise with a narrower measurable spectral range.

In order to use the setup shown in Fig. 3.18 to determine the electron temperature of, say argon glow discharge plasma (DC or RF), we first consider expression (3.23) and assume the emission to be isotropic, then the emission per steradian ($W/cm^3/sr$) is

$$I_{pq} = \frac{hcA(p,q)}{4\pi\lambda} N(p). \tag{3.36}$$

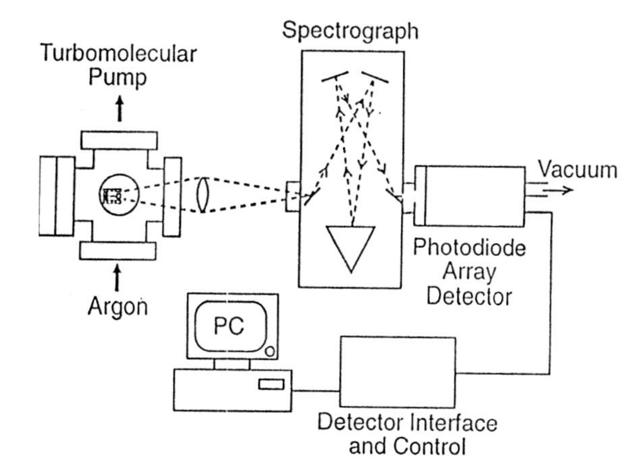

Fig. 3.18 Schematic setup of optical emission spectroscopy of plasma

Lets say we select four lines:

$$\lambda_1 = 8014.79 \text{ Å (Ar)}$$
$$\lambda_2 = 8006.16 \text{ Å (Ar}^+)$$
$$\lambda_3 = 8103.69 \text{ Å (Ar)}$$
$$\lambda_4 = 7948.18 \text{ Å (Ar}^+)$$

For line 1, with the assumptions that the plasma is at LTE and it is optically thin, we can write

$$I_1 = \frac{hcA_1}{4\pi\lambda_1} N_o^1 \tag{3.37}$$

where N_o^1 is given by expression (3.28) as $N_o^1 = N_o \frac{g_1}{U_o} \exp\left(-\frac{E_o}{kT_e}\right)$. The ratio of the intensities of line 1 to line 2 is given by

$$R_1 = \frac{I_1}{I_2} = \frac{A_1}{A_2} \frac{\lambda_2}{\lambda_1} \frac{N_o}{N_1} \frac{g_1}{g_2} \frac{U_1}{U_o} \exp\left(-\frac{E_1 - E_2}{kT_e}\right) \tag{3.38}$$

where E_1 is the energy of the upper level involved in the transition giving rise to line 1, while E_2 is that of line 2. It can be seen that in order to calculate R_1 for a particular electron temperature, we need to know the ratio of N_o/N_1 which has to be calculated from expression (3.26). Although this is possible, we may also get around this by taking the ratio of a second pair of lines:

$$R_2 = \frac{I_3}{I_4} = \frac{A_3}{A_4} \frac{\lambda_4}{\lambda_3} \frac{N_o}{N_1} \frac{g_3}{g_4} \frac{U_1}{U_o} \exp\left(-\frac{E_3 - E_4}{kT_e}\right). \tag{3.39}$$

Taking ratio of R_1 and R_2 will then give:

$$R = \frac{R_1}{R_2} = \frac{A_1}{A_2} \frac{A_4}{A_3} \frac{\lambda_2}{\lambda_1} \frac{\lambda_3}{\lambda_4} \frac{g_1}{g_2} \frac{g_4}{g_3} \exp\left(-\frac{E_1 - E_2 + E_4 - E_3}{kT_e}\right). \tag{3.40}$$

The values of R for a range of T_e can be calculated and plotted as shown in Fig. 3.19.

Experimentally, the intensity of the four lines will be measured and the value of R obtained from Eq. (3.40). The electron temperature of the plasma can then be estimated by using the "calibration curve" in Fig. 3.19.

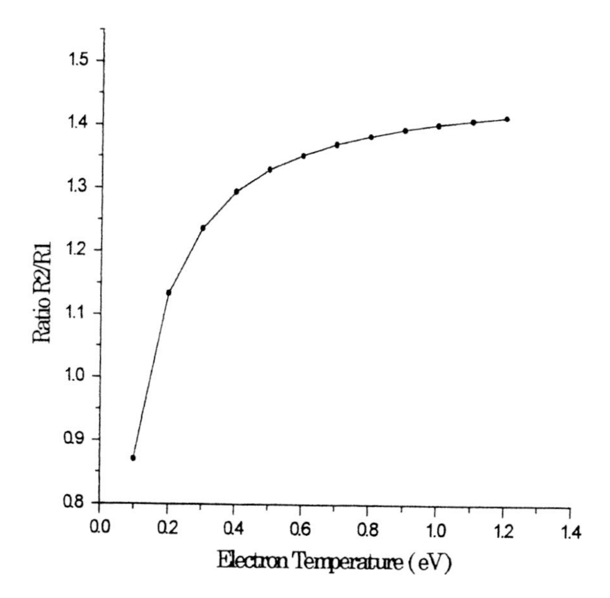

Fig. 3.19 Optical line ratio plot for electron temperature determination

An example of the spectrum of a high pressure argon spectral lamp in the region of these 4 lines is shown in Fig. 3.20.

The electron temperature of the argon plasma of the high pressure arc lamp is estimated to be 0.4 eV.

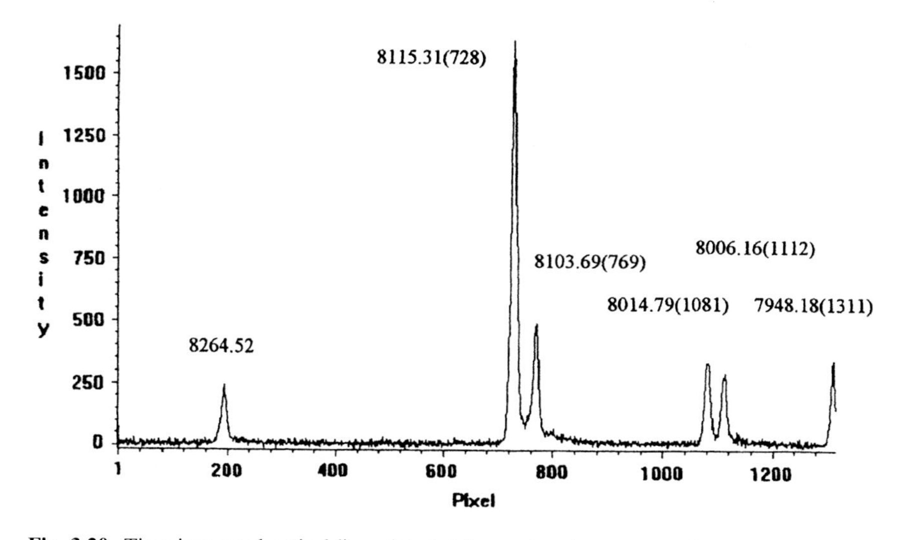

Fig. 3.20 Time integrated optical lines detected from a low temperature plasma

3.4 The Langmuir Probe (Electric Probe)

3.4.1 Electron Temperature and Density Measurements of Steady State Plasmas

The Langmuir probe, also often known as the electric probe, is one of the frequently used diagnostic tools for measuring the characteristics of low temperature plasmas. The fundamental plasma parameters, such as the electron density, electron temperature, plasma potential and in some cases the electron energy distribution function can be determined by using the Langmuir probe.

The theory of the flow of charge carriers to an electric probe is extremely complex. A simplified theory can be derived with the following assumptions:

i. The electron and ion concentrations are equal.
ii. The mean free paths of the electrons and ions are much larger than the probe dimensions.
iii. The electron temperature is much larger than the ion temperature.
iv. The probe radius is much larger than the Debye length of the plasma.
v. The kinetic energy of electrons and ions obey Maxwell-Boltzmann distribution.

The Langmuir probe can be either cylindrical, planer or spherical in shape. In its simplest form, it consists of a short length of thin wire inserted into the plasma, which is biased at a potential with respect to a reference electrode (usually the electrically grounded electrode) in order to collect electron and/or positive ion current. As the wire is in contact with the plasma, the dimension of the probe must be chosen so as to cause minimum disturbance to the plasma. The material used for the probe construction is such that they do not melt or easily sputtered so as to avoid introducing impurity to the plasma. The commonly used materials for probe construction include tungsten, molybdenum, tantalum etc. The probe wire of 0.1–1 mm diameter is embedded in an insulating tube such as alumina or quartz so that it is insulated from the plasma except for a short length of exposed tip which is 2–10 mm in length. Such a probe is capable of providing a localized measurement of the fundamental plasma parameters such as electron temperature, electron density as well as plasma potential.

A simplified schematic of a typical probe circuit and the probe *I*–*V* characteristics are shown in Figs. 3.21 and 3.22, respectively. The probe system consists of the probe, a variable power supply and an arrangement to measure the current drawn by the probe. The variable voltage source should be able to provide output voltage that can be varied from positive to negative range with respect to ground, either discretely or scan continuously. Depending on the polarity of the biasing voltage, the probe draws either positive or negative current from the plasma.

Figure 3.22 shows an idealized probe *I*–*V* characteristics labelled at various regions by A, B, C, D, and E. Along the X-axis V (or V_P) represents the probe voltage or the probe bias voltage, while along the Y-axis I represents the total probe current. The probe current is due to the flow of both ions and electrons drawn from

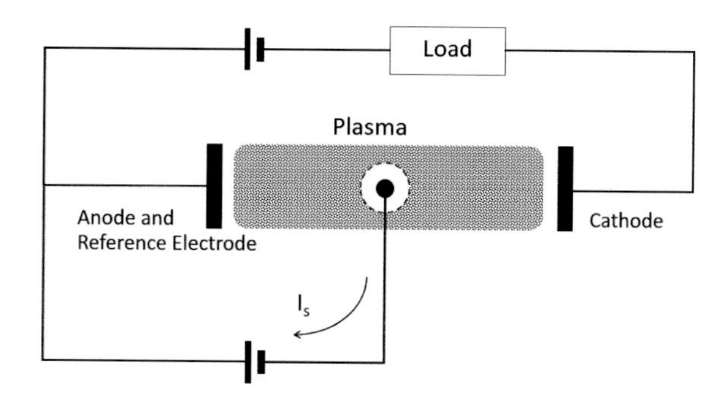

Fig. 3.21 Schematic setup of Langmuir probe measurement of a glow discharge

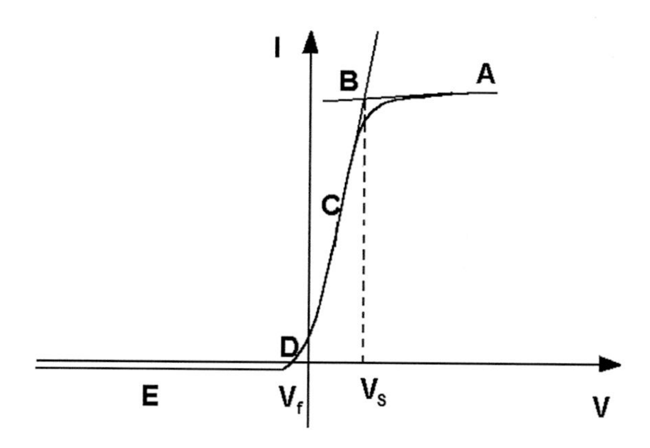

Fig. 3.22 Typical *I–V* characteristic curve of Langmuir probe (Single)

the plasma in the vicinity of the probe hence it can be expressed as $I = I_e - I_i$, where I_e is the electron current and I_i is the ion current. In Fig. 3.22, the ion current to the probe has been assigned to be negative while the electron current is assigned to be positive. Region AB is called the electron saturation region, BCD the electron retardation region and DE the ion saturation region. The floating potential of the probe is represented by V_f (point D). The floating potential V_f is the voltage at which the ion and electron currents are equal and hence the net probe current is zero. The plasma potential V_S (point B) is the potential of the plasma at the location of the probe with respect to the reference electrode.

The electron temperature of the plasma can be determined by plotting $\ln(I_e)$ against V for the electron retardation region (BCD) as shown in Fig. 3.23.

According to the probe theory, the electron current in the electron retardation region is given by

$$I_e = I_{e0}\, exp\left(\frac{eV}{kT_e}\right), \quad \text{where } I_{e0} = A\, n_e e\, \sqrt{\frac{k\,T_e}{2\pi\, m_e}}.$$

Here A is the effective probe area, n_e is the electron density, T_e is the electron temperature, while e, k and m_e are respectively the electronic charge, Boltzmann

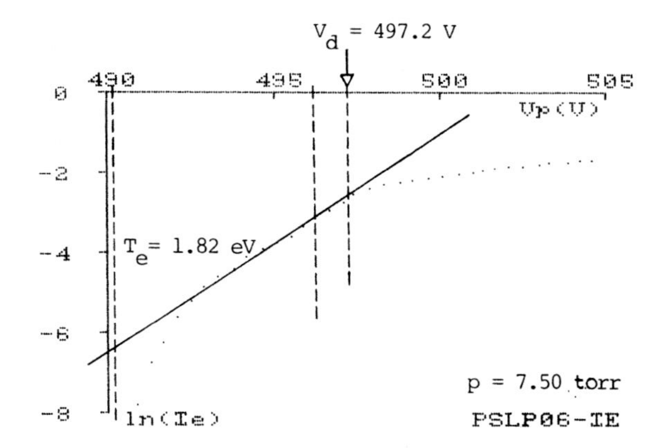

Fig. 3.23 Analysis of the *I–V* probe characteristic for the determination of electron temperature

constant and electron mass. I_{e0} is called the electron saturation current, which is the probe electron current at point B. Point B can be identified to be at the point where the gradient changes, which occurs at the point where the probe potential is V_d, which is the plasma potential V_s with corresponding current I_{e0}. The plot of $\ln(I_e)$ against V for the electron retardation region gives rise to a linear graph with gradient given by e/kT_e, hence the electron temperature T_e can be obtained. By measuring the electron saturation current I_{e0} at point B of the probe *I–V* characteristics, and knowing the probe dimension, the electron density n_e can be estimated.

3.4.2 Dynamic Studies of a Flowing Plasma (or Charged Particle Beam)

The electric probe measurement of IV characteristic and its analysis outlined in Sect. 3.4.1 can be used to measurement of electron temperature and density of a steady state plasma with low temperature and density such as the glow discharge plasma. It is assumed that the energies of the particles inside the plasma, both electrons and ions, are obeying Maxwell-Boltzmann distribution function. For a flowing plasma with time varying temperature and density, the electric probe can still be employed to study the dynamics of the plasma such as those in the shock tube. For the electromagnetic shock tube, due to the presence of the electric and magnetic field, the probe signal may become complicated. However, for a free flowing plasma consisting of charged particles, or a beam of charged particles (ions or electrons) flowing in a region without electric and magnetic fields, the electric probe can be used to detect the movement of the flowing plasma or charged particles.

An example of using the electric probe for the detection of ion beam from the plasma focus (Refer to Sect. 4.2, Chap. 4) is illustrated in Fig. 3.24 [2]. Two identical probes are placed along the path of the ion beam and each is biased at a suitable negative potential, say −100 V. The probes are expected to attract the

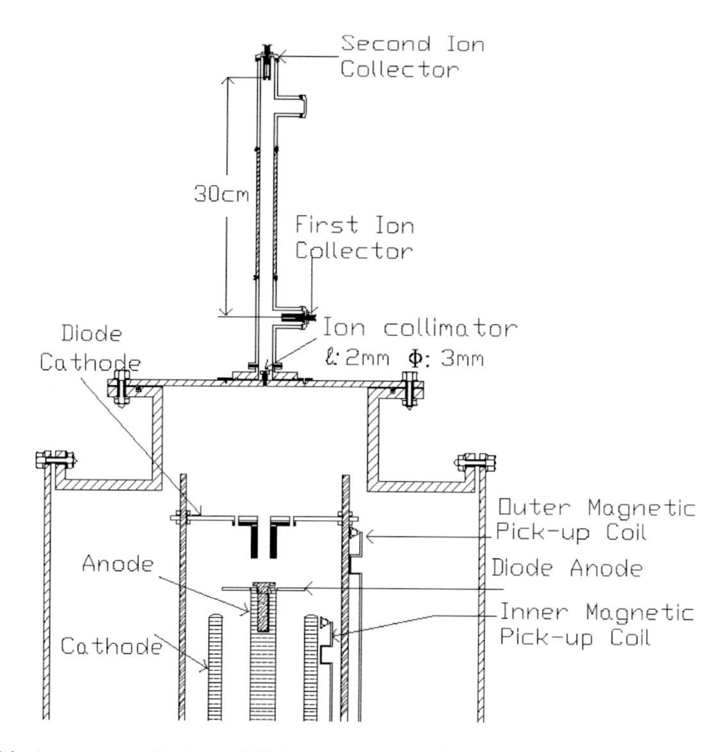

Fig. 3.24 Arrangement for time-of-flight measurement of ion beam energy using electric probes (Copyright 2002 The Japan Society of Applied Physics)

positively charged ions when the beam passes through the probes and produce two voltage pulses recorded by the oscilloscope. The probe bias circuit used is as shown in Fig. 3.25. By positioning the probes at a fixed distance apart along the path of the ion beam, the time taken by the beam to travel from one probe to another (Δt) can be measured using the oscilloscope, and hence the kinetic energy of the ions can be estimated. This is illustrated in Fig. 3.26.

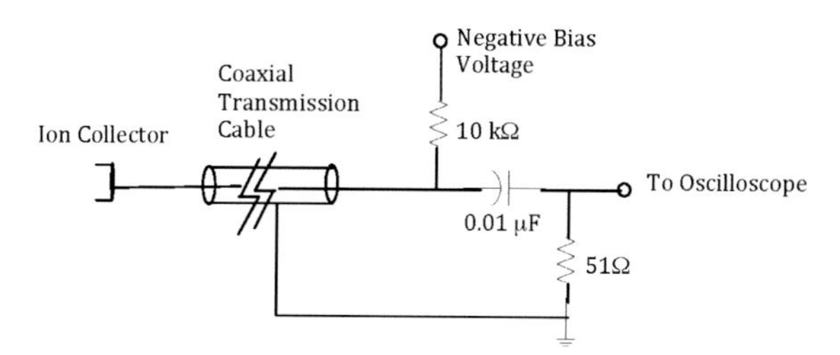

Fig. 3.25 The schematic biasing circuit for electric probe used for dynamic study of pulsed ion beam

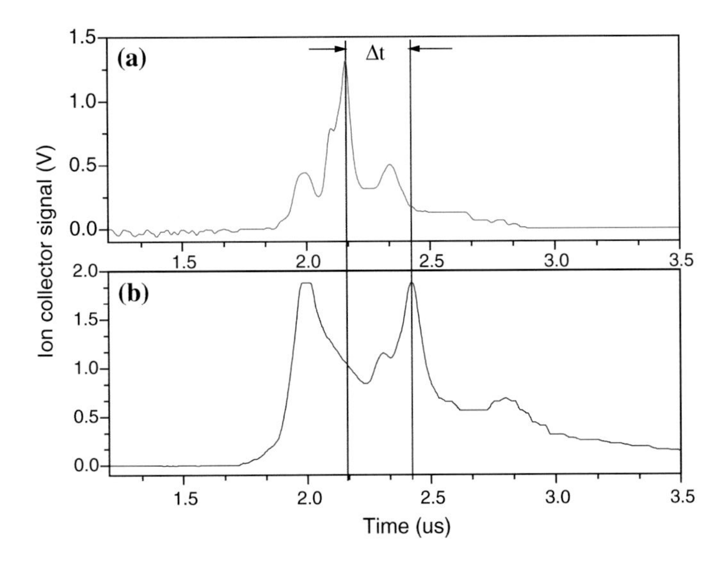

Fig. 3.26 Typical signals of electric probe measurement of ion beam

3.5 X-ray Diagnostic Techniques

As we have seen in the previous section, the plasma emission spectrum will shift towards higher photon energy when the electron temperature is increased. At an electron temperature of higher than 1 keV, the plasma emission is predominantly in the X-ray region, typically in the wavelength range of 1–100 Å. Hence for high temperature and high density plasma produced by pulsed discharge such as the plasma focus and the vacuum spark, the measurement of the X-ray emission will provide useful information concerning the condition of the plasma. In addition, in consideration of possible application of these pulsed plasmas as X-ray sources, it is important to determine its X-ray emission characteristics such as the intensity, the source size as well as the spectrum.

Figure 3.27 shows the typical arrangement for measuring the X-ray emission from a plasma source.

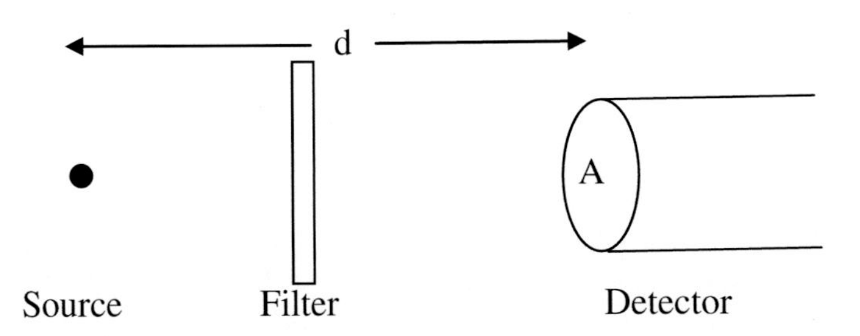

Fig. 3.27 Illustration of experimental layout for X-ray detection of plasma

In this arrangement, the detector used is spectral integrated and the filter will provide exponential filtering. In this case, the X-ray is considered to be generated from the plasma by Bremsstrahlung and recombination discussed in the earlier sections. The contribution from the line radiation will be insignificant when we integrate over the whole spectrum.

3.5.1 *X-ray Absorption Filter*

The interaction of X-ray photon with the filter material is basically the photoelectric effect where the atom absorbs the photon energy and results in the removal of one of the K-shell electrons. Some of the photons may thus be stopped during their passage through the filter. For lower energy photon (longer wavelength), a limit will be reached where the photon energy is not enough to remove the K-shell electron. This is called the K-absorption edge. At further lower photon energy, another limit will be reached for the L-shell electrons. The K-absorption edge is closely related to the ionization energy of the electron in the K-shell, which can be taken to be approximately equal to $13.6 \times Z^2$ eV considering the K-shell ionization to be "hydrogen-like" and ignoring the screening effect due to the other electrons (which will cause a lowering of the ionization energy).

The absorption of photons passing through a particular filter can be expressed as

$$I = I_o \exp(-\mu x) \tag{3.41}$$

where I is the intensity of X-ray after passing through the filter, I_o is the X-ray intensity before passing through the filter, x is the thickness of the filter in cm and μ is the absorption coefficient in cm^{-1}. A more commonly used form of the absorption coefficient is the mass absorption coefficient μ_{m} (cm^2/g) which is equal to μ/ρ where ρ is the density of the filter material in g/cm^3. In this case expression (3.41) will be written as

$$I = I_o \exp(-\mu_m x_m) \tag{3.42}$$

$x_m = \rho x$ is called the mass thickness expressed in g/cm^2. The values of μ_m for various materials are determined experimentally only for certain limited and scattered ranges of wavelength. Their values at other wavelengths are either obtained by interpolation and extrapolation, or by using some empirical expressions. For filter of composite material $X_m Y_n$, the resultant X-ray absorption coefficient is written as

$$\mu_{xy} = \left(\frac{m}{m+n}\right)\mu_x + \left(\frac{n}{m+n}\right)\mu_y. \tag{3.43}$$

As an example, the mass absorption coefficient of mylar ($C_{10}H_8O_4$) as a function of wavelength is shown in Fig. 3.28.

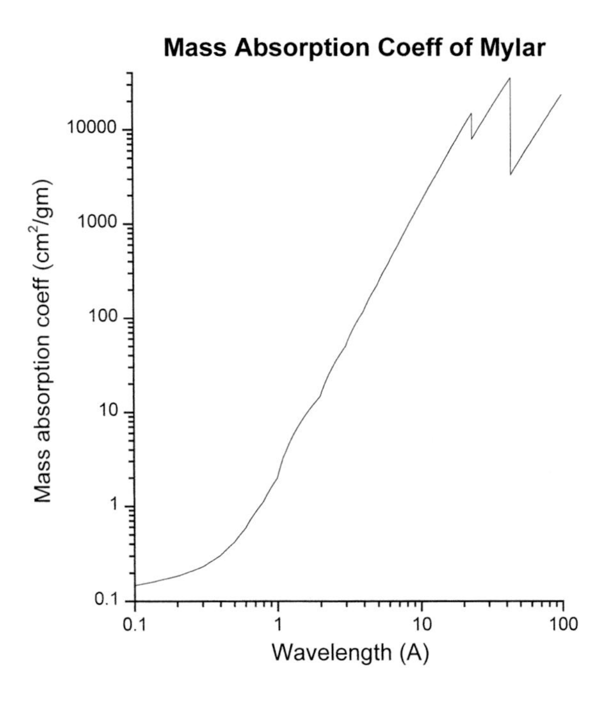

Fig. 3.28 Mass absorption coefficient of Mylar as a function of X-ray wavelength

The X-ray transmission through various sets of filters are illustrated graphically by plotting expression (3.42) with known values of μ_m as shown in Fig. 3.29. In particular, we note the following three sets of filters: (1) 149 μm aluminized-Mylar; (2) 24 μm aluminized-Mylar + 10.5 μm Ti; and (3) 1024 μm aluminized-Mylar. If these filters are used simultaneously with identical detectors, their signals can be used to identify the presence of argon K_α line radiation in an argon plasma or an argon-hydrogen admixture plasma, for example. This is due to the fact that the argon K_α line radiation which has a wavelength of around 4 Å, will be within the transmission "window" of Ti. With the above sets of filters, the signals registered by the detector with filters (1) and (2) should be almost identical if the argon K_α line radiation is dominant. On the other hand, if copper K_α line radiation is prominent, the signals of detectors with filters (2) and (3) will be identically weak while that of detector with filter (1) will be strong. It should also be noted that all materials will transmit its own K_α line radiation well.

3.5.2 X-ray Detectors

3.5.2.1 Time-Resolved X-ray Detectors

It is very useful to obtain information concerning the time-evolution of the X-ray emission from the plasma as this can often reveal the dynamic behaviour of the

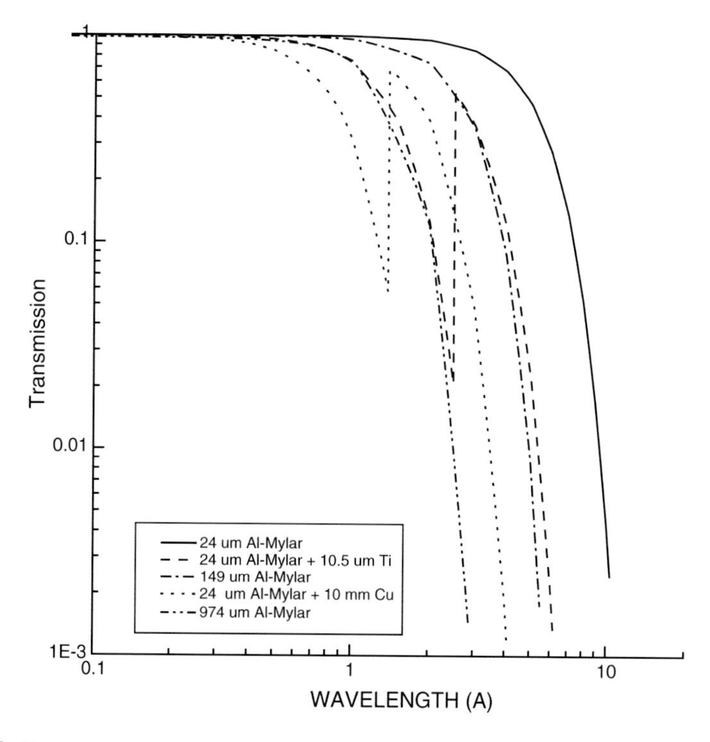

Fig. 3.29 X-ray transmission curves for various sets of absorbers

plasma. Several common X-ray detectors used for this purpose include the photomultiplier tube, silicon PIN diode and the X-ray diode (XRD).

(a) The photomultiplier tube

The photomultiplier tube (PMT) is a sensitive photon detector which is capable of converting photons into electrons using a photo-cathode and subsequently multiply them through multiple stages of dynode arrangement. An example of the photomultiplier tube is shown in Fig. 3.30. The PMT requires a "base" which normally

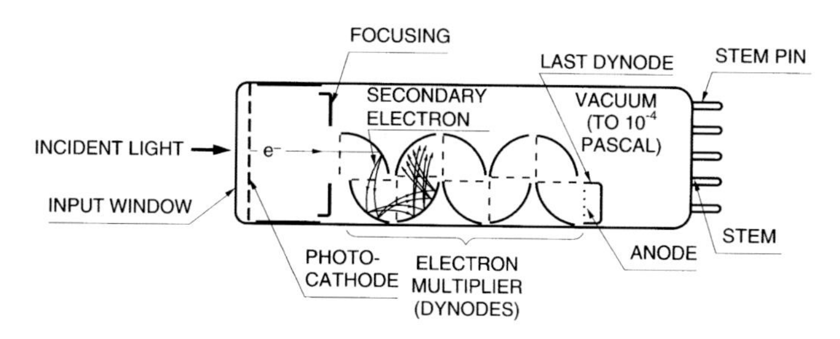

Fig. 3.30 Schematics of a typical photomultiplier tube

consists of a resistive divider network to provide a distribution of potential across the cathode and the first dynode as well as between consecutive dynodes. The potential differences between the cathode and the first dynode provides the first acceleration of the photoelectrons and subsequently an almost constant potential is maintained between two consecutive dynodes until finally the electrons are collected by the anode.

For the detection of X-ray, the PMT is used together with a scintillator. This is because normally the photocathode is sensitive to photon in the visible to UV region. The scintillator is capable of absorbing photon (X- and γ-rays) or particles (α, β and neutron) of sufficient energy and emit in the visible region (blue to violet). For X-ray, it converts photons with wavelength in the region of 1–100 Å to around 4000 Å which is within the sensitivity region of the PMT's photocathode.

As we can see, the detection of X-ray using PMT involves several steps of conversion. First the X-ray photon is absorbed by the scintillator which gives out blue-violet photon (conversion factor ε_s). Second, the quantum efficiency (ε_q) of the photocathode should be considered. Thirdly, the electron multiplication factors of the PMT (ε_m). And finally, the electrons are collected by the anode and the signal obtained (voltage) is caused by the flow of these electrons in the circuit (current). It is hence difficult, though not impossible, to calculate the overall sensitivity of the whole detection system.

Furthermore, due to the transit time required for the electron multiplication process inside the PMT, its signal is always delayed by a duration of the order of 50–200 ns, depending on the number of multiplication stages. More stages means higher gain but longer time delay. A delay of around 100 ns can be expected for a 9 stages PMT.

However, despite the fact that the PMT system is normally bulky in size, produces a delayed signal and it is difficult to determine its absolute response, it is useful if we need to measure the harder component of the X-ray emission spectrum. Figure 3.31 shows an example of the hard X-ray signal (lower trace) of a plasma focus discharge recorded by a PMT system [3]. The upper trace shows the voltage spike which is the signature of the focussing action. The timing of the PMT signal has to be adjusted to align with the voltage signal.

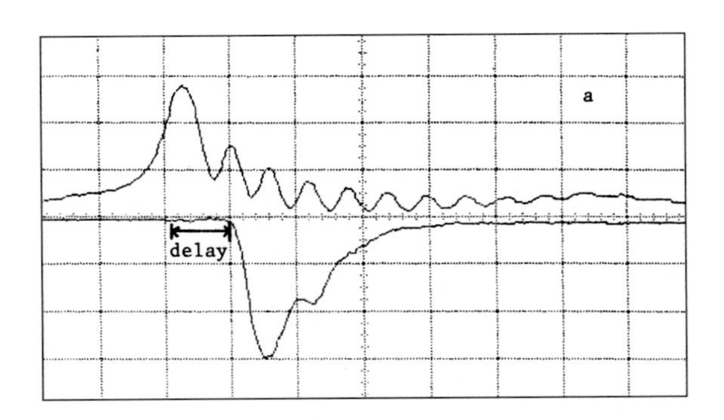

Fig. 3.31 Example of the PMT signal obtained from a plasma focus discharge

(b) Silicon PIN diode

The silicon PIN diode is a convenient X-ray detector which is compact, fast rise time and with known response. It has an almost flat response (except for the silicon absorption edge) between 1 and 20 Å. The PIN diode is basically similar to a PN junction diode, but with a large intrinsic layer sandwiched between the P-type and the N-type layer. This thick intrinsic layer is completely depleted and it will absorb the X-ray photon to form an electron-hole pair.

The structure of the PIN diode is shown schematically in Fig. 3.32 together with the energy level diagram when a reverse bias voltage is applied across the diode. Electron-hole pairs will be produced when X-ray photons are absorbed by the diode, with a response of 1 electron-hole pair produced for each 3.55 eV of photon energy absorbed. This means electron charge of 0.28 C will be produced for each joule of X-ray energy absorbed.

To calculate the sensitivity of the diode to X-ray photons, the absorption of X-ray at the N-type entrance window which acts as filter must be considered. The electron-hole pair formation will correspond to the amount of photons absorbed by the intrinsic layer and hence the sensitivity $S(\lambda)$ of the diode is given by

$$S(\lambda) = 0.28 \exp(-\mu_s x_1)[1 - \exp(-\mu_s x_2)] \quad (\mathrm{C/J}) \tag{3.44}$$

where μ_s is the mass absorption coefficient of silicon, x_1 is the mass thickness of the N-type entrance window while x_2 is the mass thickness of the intrinsic layer. An example of the sensitivity curve of a typical X-ray PIN diode is shown in Fig. 3.33.

At the short wavelength region ($\lambda < 1$ Å), the X-ray photons may pass through the intrinsic layer without being absorbed so the sensitivity drops. On the other hand, for long wavelength region ($\lambda > 20$ Å), the X-ray photon will be absorbed by the silicon entrance layer and hence the sensitivity will also drop. In the region between this two limits, the sensitivity of the diode is almost flat, except at the K-absorption edge of Si at 6.73 Å.

The X-ray intensity from the plasma source $P(\lambda, T_e)$ W cm^{-3} Å^{-1} actually detected by the PIN diode with due consideration of the absorption by foil placed between the source and the detector and the solid angle (assuming point source) is then given by $P(\lambda, T_e)S(\lambda)\exp(-\mu x)$. Figure 3.34 show the example of the intensity of X-ray emitted by an argon plasma at electron temperature of 1 keV detected by a PIN diode covered with various sets of filters.

A commonly used biasing circuit for the PIN diode is as shown in Fig. 3.35.

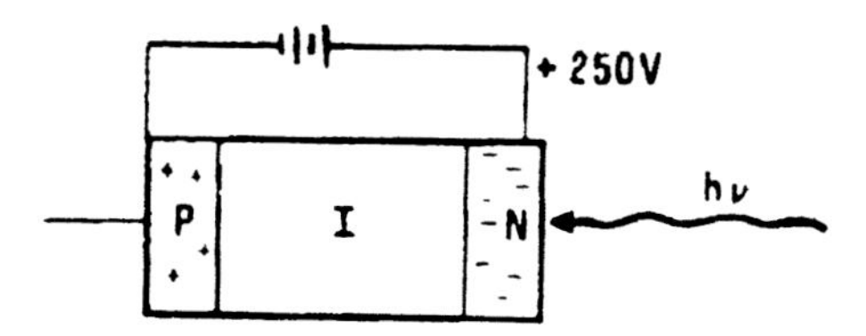

Fig. 3.32 Simplified schematics of a PIN diode

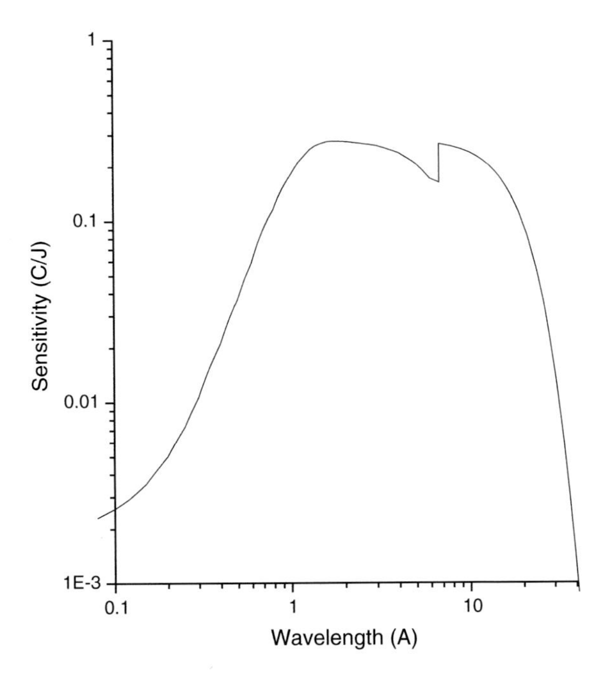

Fig. 3.33 Theoretical sensitivity curve of an X-ray PIN diode

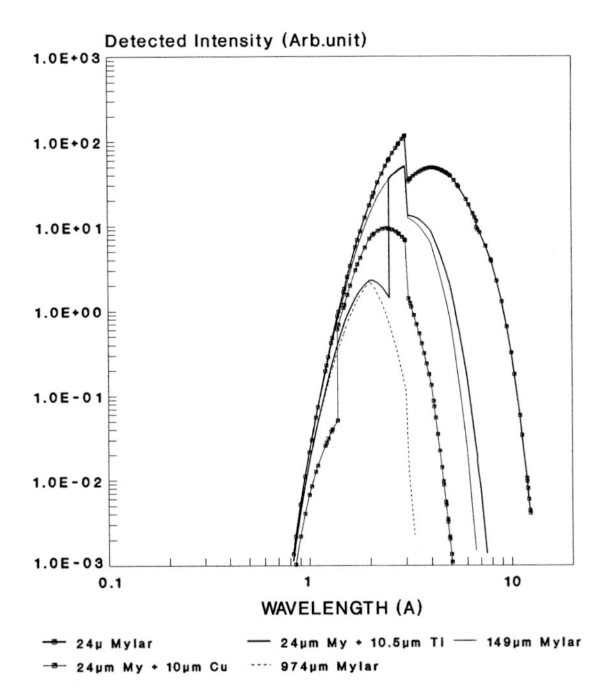

Fig. 3.34 Calculated X-ray spectra detected by PIN diode after various sets of filters

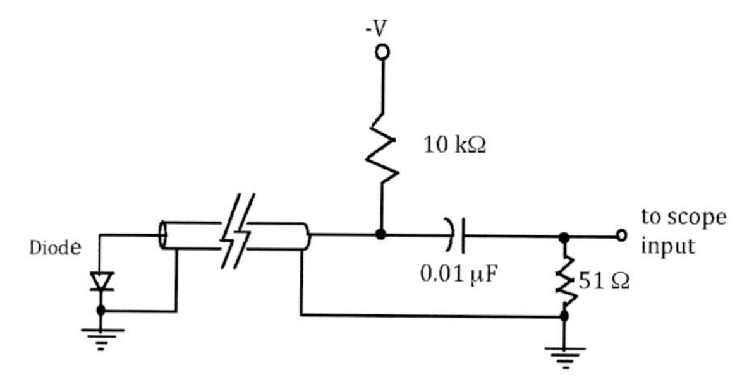

Fig. 3.35 Schematics of a typical PIN diode biasing circuit

In this circuit, the charges produced by the diode due to the absorption of X-ray energy will flow through the circuit and the output voltage is measured across the 51 Ω resistor.

An example of the X-ray pulse measured is as shown in Fig. 3.36. The amount of charges produced by the PIN diode due to the absorption of X-ray photons is given by:

$$Q = \int I(t)dt = \frac{\int V(t)dt}{51}.$$

If the source is a monochromatic point source with wavelength $\lambda = \lambda_o$ (line radiation), then the charge Q can be used to calculate the amount of X-ray energy absorbed:

$$E''_{\lambda_o} = \frac{Q}{S(\lambda_o)}.$$

Corrected for foil absorption:

$$E'_{\lambda_o} = \frac{Q}{S(\lambda_o)} \cdot \frac{1}{\exp(-\mu x)}$$

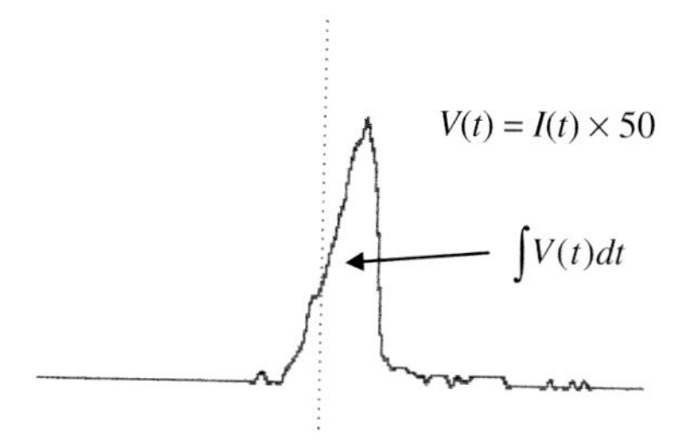

Fig. 3.36 Example of a PIN diode X-ray pulse registered by the oscilloscope

and finally, the total X-ray energy emitted by the point source isotropically into 4π is:

$$E = \frac{Q}{S(\lambda_o)} \cdot \frac{1}{\exp(-\mu x)} \cdot \frac{4\pi d^2}{A} \tag{3.45}$$

where d is the source-detector distance, and A is the effective area of detection of the diode as shown in Fig. 3.27.

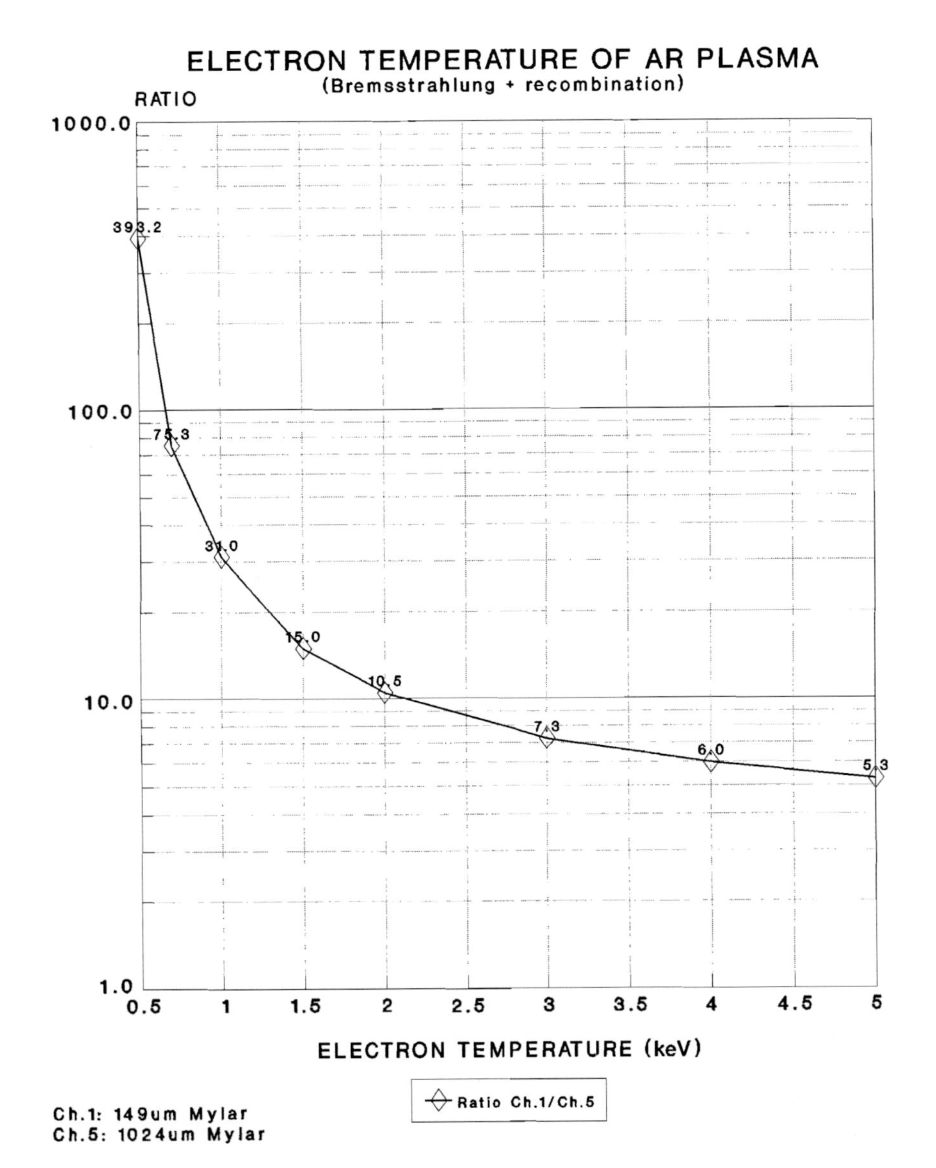

Fig. 3.37 The ratios of a pair of identical PIN diode filtered by two sets of filter as a function of electron temperature

For plasmas such as the plasma focus and the vacuum spark, it is clear that the spectrum is not monochromatic. If $d \gg$ dimension of the source which is of the order of mm (end-on) for the focus and a fraction of mm for the vacuum spark, the point source assumption may still be maintained. For such cases, we consider the continuum (Bremsstrahlung + recombination) and we write the rate of charge production (current) due to the time varying X-ray pulse emitted from the plasma:

$$\frac{dQ}{dt} = \frac{V(t)}{51} = V_{plasma} \int_{all\ \lambda} P(\lambda, T_e) \cdot \frac{A}{4\pi d^2} \cdot \exp(-\mu x) \cdot S(\lambda) d\lambda. \qquad (3.46)$$

This can be computed for a fixed T_e. For two identical diodes (or any other detectors) coupled to two different sets of foils, the ratio of the current generated by the two diodes due to the same X-ray pulse can then be calculated as $R = I_1/I_2$. This can be repeated for a range of temperature, say from 500 eV up to 10 keV, the values of R as a function of T_e can be obtained and plotted as shown in Fig. 3.37. This can be used as calibration curve for the measurement of electron temperature by the X-ray ratio method or X-ray foil absorption technique. For this case, two channels of PIN diode are required.

Alternatively, R for filters of same material (e.g. Al) but varying thickness can be computed and plotted as R against foil thickness for a particular T_e. This can be repeated for a range of temperature. The set of graphs for T_e = 500, 750, 1000, 2000, 5000 and 10,000 eV are as shown in Fig. 3.38. The graph for monochromatic sources of Ar K_α and Cu K_α are also shown in this figure.

Experimentally, this requires more than 2 channels of detectors. For the case of 5 channels of detectors, with one channel as reference, 4 points can be obtained experimentally and plotted superimposed onto the ratio graphs in Fig. 3.38. The experiment is carried out on a plasma focus discharge and the signals of the five channels PIN diodes are shown together with the voltage signal as shown in Fig. 3.39 [4].

From this experiment, five X-ray pulses are observed, of which pulse 1 and pulse 2 can be identified to be corresponding to hot plasmas with electron temperatures between 1 and 2 keV. Pulse 4 and pulse 5 which occurs at time long after the focus event is over, are found to be close to the Cu K_α line indicating that they consist of predominantly the Cu Kα line radiation. These are believed to be originated from dense (but not so hot) plasma jet of plasma vapour.

3.5.2.2 Time-Integrated X-ray Imaging

The time-resolved measurement of X-ray emission from plasmas described above provides information concerning the time evolution of plasma but it does not tell us information such as the size and the structure of the X-ray emitting plasma. Such kind of information must be obtained by imaging the plasma. A very simple setup

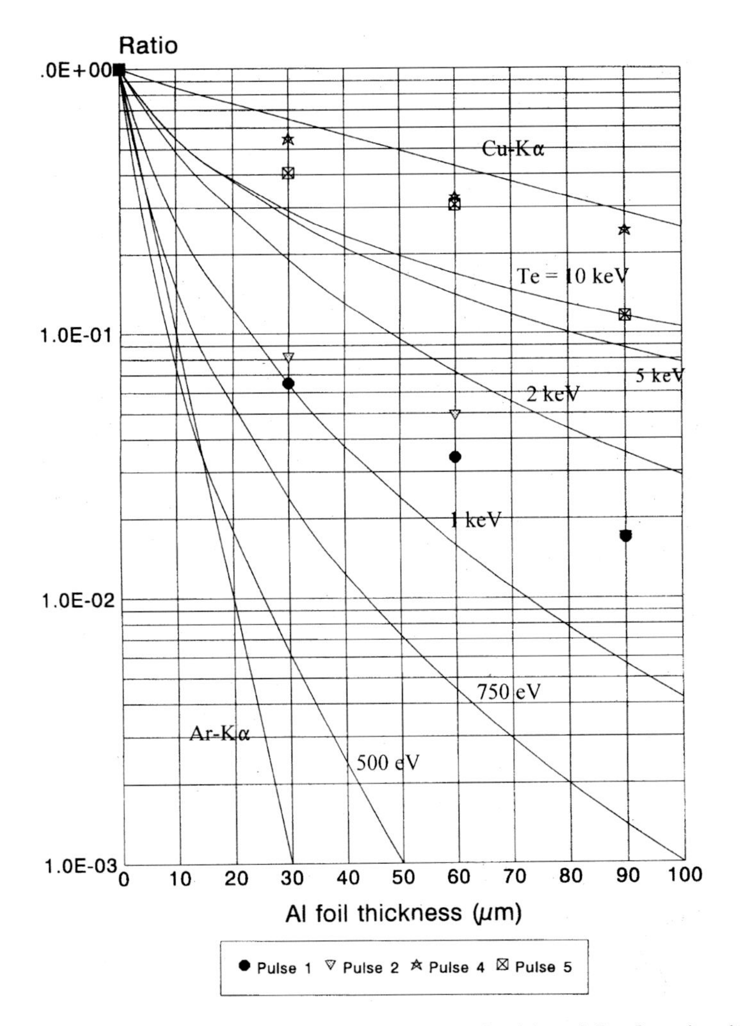

Fig. 3.38 The ratios of a series of PIN diodes filtered by aluminium foils of varying thickness plotted for various electron temperatures. Also plotted on the same graph are the *straight line curves* corresponding to mono energetic photons of Ar K_α and Cu K_α line radiations

using time-integrated X-ray film and the primitive pinhole imaging method can be used to obtain the image of the plasma. Recently, with the availability of gated micro-channel-plate (MCP), time-resolved image recording can be achieved when coupled to well designed multiple pinhole system. However, we will illustrate here only the basic technique of time-integrated X-ray imaging, that is the X-ray pinhole imaging using X-ray film.

The principle of normal pinhole imaging is straight forward. In this case, the size of the pinhole is much smaller than that of the object and the image to object ratio is given by the image-pinhole distance to object-pinhole distance ratio. However, if

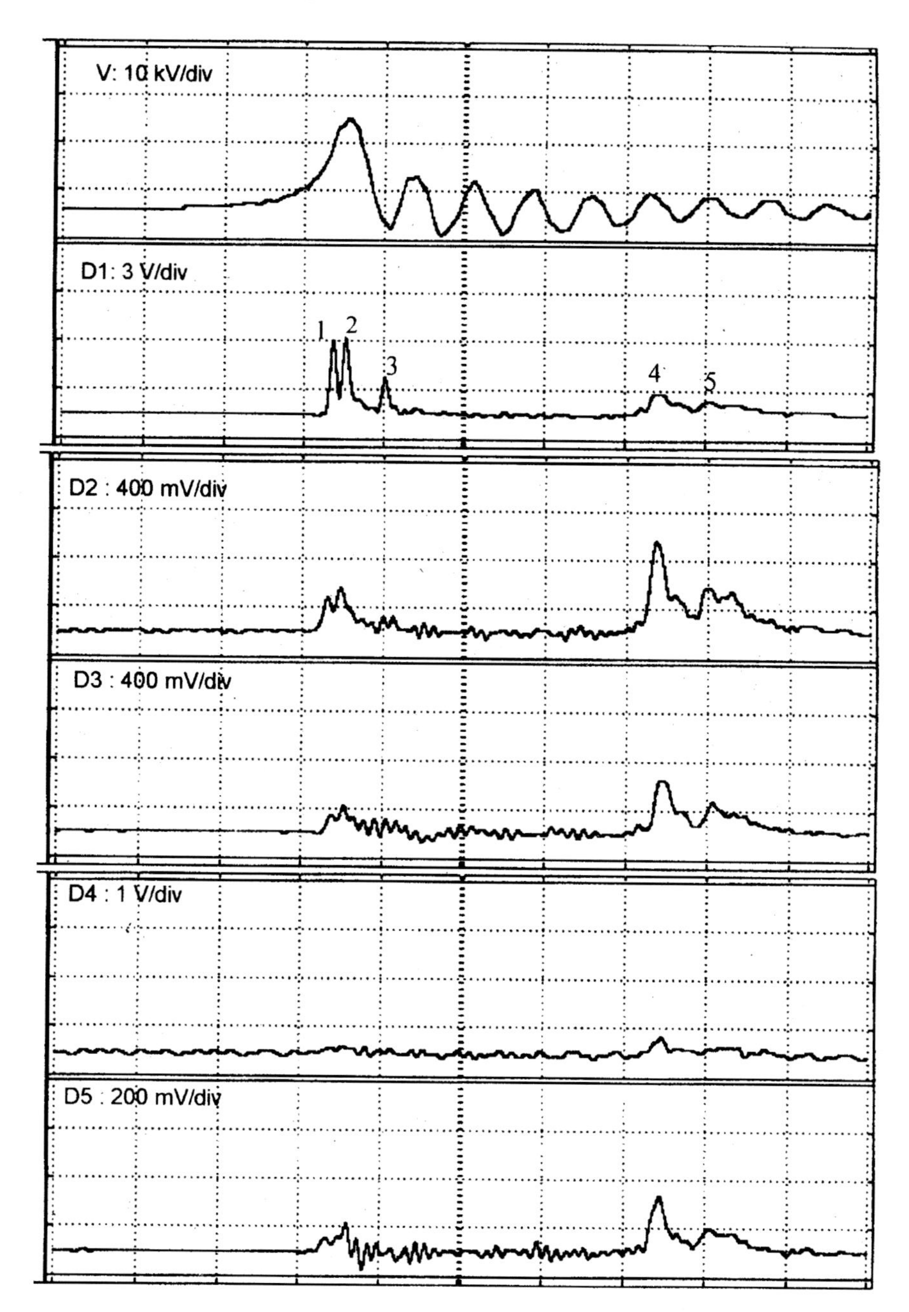

Fig. 3.39 Example of X-ray pulses from 5 channels of PIN diode filtered by aluminium foils of varying thicknesses corresponding to a single plasma focus discharge

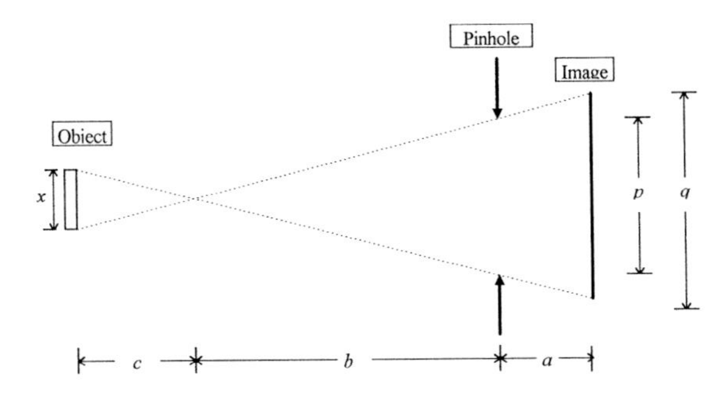

Fig. 3.40 Schematics of an X-ray pinhole imaging setup with pinhole size larger than the image size

the size of the object is smaller than the size of the pinhole, the ray diagram is as shown in Fig. 3.40.

From the geometry, we see that the size of the object can be estimated from

$$x = \left(\frac{L_o(q-p)}{L_i}\right) - p \quad \text{for } x < p$$

where L_i is the image-pinhole distance (=a) and L_o is the object-pinhole distance (=b + c). If $x > p$, we simply have $x = \left(\frac{L_o}{L_i}\right)q$.

An example of the X-ray pinhole image of the vacuum spark plasma is shown in Fig. 3.41.

In this image, we can clearly see 2 types of structures: (i) a bright spot with sharp circular (almost) boundary, and (ii) a diffuse cloud of intensity. The bright round spot is believed to be caused by the commonly observed hot spot formed in the vacuum spark plasma, which is smaller than the size of the pinhole which has a diameter of 300 μm. In fact, the sharp circular structure suggests that it is actually the projection of the pinhole onto the X-ray film by a source much smaller than the pinhole size. On the other hand, the plasma cloud is much larger in dimension than 300 μm so this part of the image shows the actual structure of the plasma cloud observed in the vacuum spark.

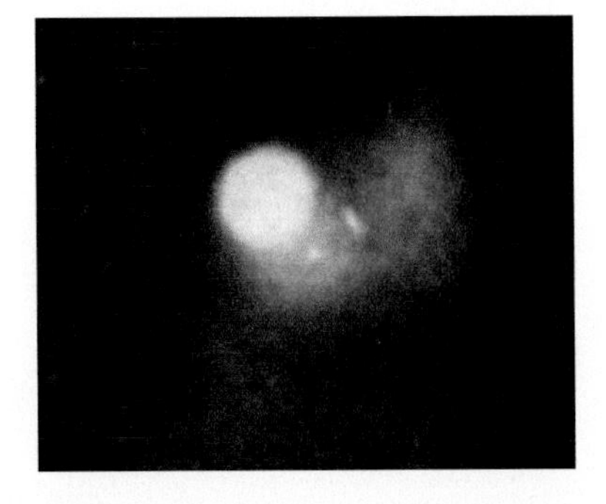

Fig. 3.41 Example of the X-ray pinhole image obtained from a vacuum spark discharge

3.6 Neutron Diagnostic Techniques

3.6.1 *The Foil Activation Neutron Detector for Absolute Neutron Yield Measurement*

Many nuclides such as ^{115}In, ^{107}Ag, ^{109}Ag and ^{103}Rh have large cross section for the (n, γ) reaction with thermal neutron (<1 keV). The products of such a reaction are the gamma ray and the radioactive isotope of the target nuclide. By measuring the activity of the radioactive isotope for a few half-lives of its decay, the incident neutron flux can be deduced. The activation foil used is in the form of a thin foil with less than 1 mm thickness so as to reduce the self absorption of the radiation by the material.

Consider a piece of foil consisting of N_T number of nuclides. When this foil is exposed to thermal neutrons of flux density ϕ with its plane perpendicular to the direction of the neutrons, the rate of change of induced radionuclides in the foil at any instant is given by:

$$\frac{dN}{dt} = N_T \sigma_a \phi - \lambda N \tag{3.48}$$

where

N is the number of radionuclides at any instant,

σ_a is the average thermal neutron activation cross section,

λ is the decay constant of the radionuclides.

The first term on the right hand side of Eq. (3.48) is the activation term while the second term represents the decay of the activated radionuclides which occur simultaneously with the activation process. The number of radionuclides in the foil after activation for a time of t can be obtained by solving Eq. (3.48). The solution takes the form:

$$N(t) = \frac{N_T \sigma_a \phi}{\lambda}\left(1 - e^{-\lambda t}\right). \tag{3.49}$$

This can be represented graphically as shown in Fig. 3.42.

If t is sufficiently large, the value of $N(t)$ approaches a constant value known as the saturation activity given by

$$N_s = \frac{N_T \sigma_a \phi}{\lambda}. \tag{3.50}$$

Practically, more than 95 % of this saturation value is achieved by activating the foil for a time of about five half-lives.

We consider neutron activation by a continuous source and a pulsed source.

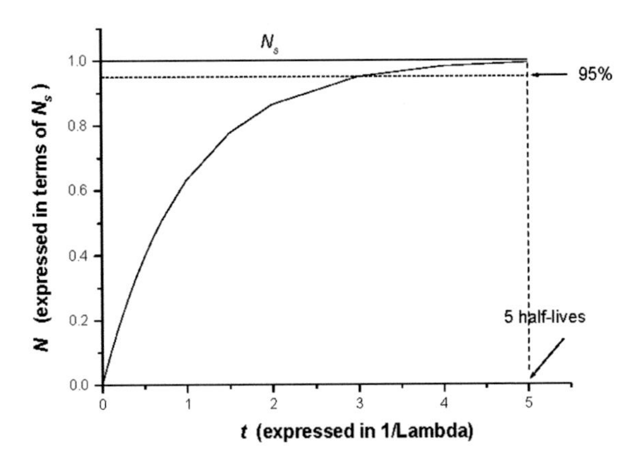

Fig. 3.42 Variation of number of neutron activated radionuclides with exposure time

3.6.1.1 Steady Source

For a steady and isotropic neutron source with strength S neutrons per second placed at a fixed distance from the foil, the neutron flux density at the surface of the foil can be expressed as:

$$\phi = \frac{f\,S}{a} \tag{3.51}$$

where

f is the fraction of the emitted neutrons reaching the foil, and
a is the area of the foil.

The factor f takes into account the scattering and absorption of the neutrons in the absorber placed in between the source and the foil (the atmospheric air and the paraffin block used as moderator) as well as the geometrical factor (solid angle consideration). Substituting for ϕ in Eq. (3.50) gives:

$$N(t) = \frac{N_T \sigma_a f S}{\lambda a}\left(1 - \mathrm{e}^{-\lambda t}\right) \tag{3.52}$$

and the activity of the radionuclides after an activation time of t_a is

$$A_o = \lambda N = \frac{N_T \sigma_a f S}{a}\left(1 - \mathrm{e}^{-\lambda t_a}\right). \tag{3.53}$$

If the activity of the foil is measured by a counting system of efficiency ε for a period of t_i immediately after activation, the total number of counts obtained is given by

$$C = \varepsilon \int_0^{t_i} A_o \exp(-\lambda t) dt$$
$$= \left(\frac{\varepsilon f \, N_T \sigma_a}{a} \right) \frac{S}{\lambda} \left(1 - e^{-\lambda t_a}\right)\left(1 - e^{-\lambda t_i}\right). \tag{3.54}$$

This equation relates the total counts to the strength of the continuous neutron source. It can be written in the form $C = KS$ if the activation time t_a and the counting time t_i are fixed. K is then the calibration constant of the detector setup that may be calculated from

$$K = \left(\frac{\varepsilon f N_T \sigma_a}{a\lambda} \right) \left(1 - e^{-\lambda t_a}\right)\left(1 - e^{-\lambda t_i}\right) \tag{3.55}$$

provided all the constants can be known with sufficient accuracy. Alternatively, K can be determined by calibrating the detector against a source with known strength as

$$K = \frac{C}{S}.$$

3.6.1.2 Pulsed Neutron Source

For a pulsed source, the equivalent source strength may be specified as total yield (n) divided by the pulse duration (τ) assuming square pulse, and the activation time will be equal to the pulse duration. Hence expression (3.54) will be re-written as

$$C' = \left(\frac{\varepsilon f N_T \sigma_a}{a} \right) \frac{n}{\tau\lambda} \left(1 - e^{-\lambda \tau}\right)\left(1 - e^{-\lambda t_i}\right). \tag{3.56}$$

Since

$$\left(\frac{1 - e^{-\lambda \tau}}{\tau} \right) \xrightarrow[\tau \to 0]{} \lambda,$$

we get

$$C' = \left(\frac{\varepsilon f N_T \sigma_a}{a} \right) \frac{n}{\lambda} \left(1 - e^{-\lambda t_i}\right) \tag{3.57}$$

which relate the counts obtained for exposure of the detector to a pulsed source of yield n. Comparing (3.57) with (3.54), we can obtain

$$n = \frac{S}{\lambda}\left(1 - e^{-\lambda t_a}\right)\frac{C'}{C}. \tag{3.58}$$

Hence the foil activation neutron detector can be calibrated against a constant source as standard and then used to measure the yield of a pulsed neutron source provided the same counting time is used for both the calibration and the pulsed neutron measurement. The factor

$$F = \frac{S}{\lambda}\left(1 - e^{-\lambda t_a}\right)\frac{1}{C}$$

is the calibration factor of the detection system obtained from the calibration against a constant source of strength S with C as the counts obtained for a counting time of t_i.

An example of the material that may be used for the detection of pulsed neutron source such as the plasma focus is indium. The activation and decay schemes of indium are as shown in Fig. 3.43.

Natural indium consists of 95.8 % of ^{115}In and 4.2 % of ^{113}In. During activation, both components will be activated. However, the activation cross section of ^{113}In is insignificantly small hence its contribution will be neglected. There are two possible branches of activation of ^{115}In by thermal neutrons: one has a cross section of about 160 barns and the product radionuclide is ^{116m}In. Another branch is that of ^{116}In which has a cross section of about 42 barns, producing ^{116}In. Both these radionuclides undergo decay to the stable nuclide ^{116}Sn.

Activation schemes: $^{115}\mathrm{In} + {}^{1}\mathrm{n} \rightarrow {}^{116m}\mathrm{In} + \gamma \quad (\sigma_a = 160 \pm 2 \text{ barns})$

$^{115}\mathrm{In} + {}^{1}\mathrm{n} \rightarrow {}^{116}\mathrm{In} + \gamma \quad (\sigma_a = 42 \pm 1 \text{ barns})$

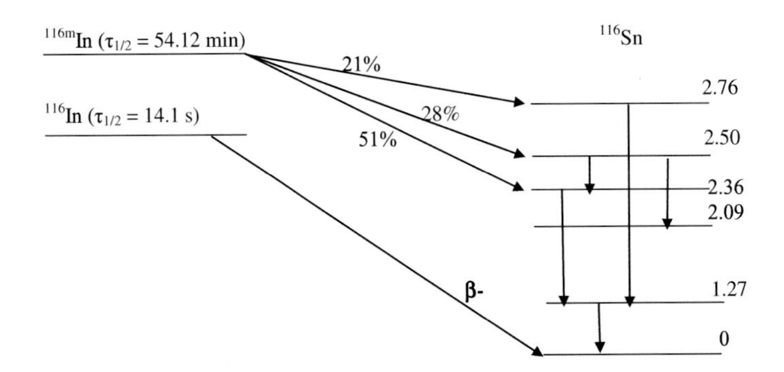

Fig. 3.43 Neutron activation and decay scheme of indium

The presence of two branches of decay implies that their contribution to the radiation activity of the activated indium foil must be considered together. However, if we choose the activation time and the counting time to be much shorter than the half-life of the ^{116m}In branch ($\tau_{1/2}$ = 54.12 min), its contribution can be made relatively insignificant compared to that of ^{116}In. This will be the case when we choose t_a to be 70 s (~5 half-lives of ^{116}In's decay) and t_i to be 40 s (~3 half-lives of ^{116}In's decay). Hence for the Indium Foil Neutron Activation Detector (IFNAD), we may obtain

$$F = 19.4\left(\frac{S}{C}\right)$$

where S is the strength of the standard neutron source used in the calibration and C is the counts obtained in 40 s immediately after activation for 70 s. If the counts obtained for the pulsed source in 40 s is C', then the neutron yield of that pulse is

$$n = F \times C'.$$

Note that C' must be counts obtained for the same counting time as C, which is the counts obtained against the standard calibration source.

3.6.2 The Detector Setup

One possible setup of the neutron foil activation detector is shown schematically in Fig. 3.44. It consists of an activation foil (indium) attached directly to a plastic scintillator coupled to a PMT. The PMT is usually mounted onto a base consisting

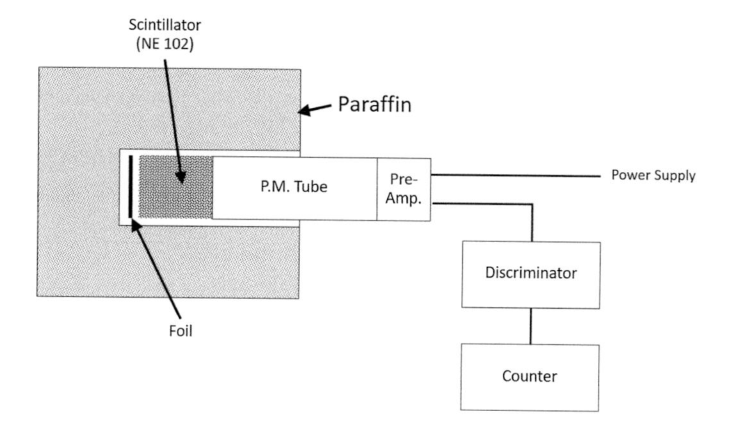

Fig. 3.44 Schematic setup of a neutron activation detector using photomultiplier tube as the detector

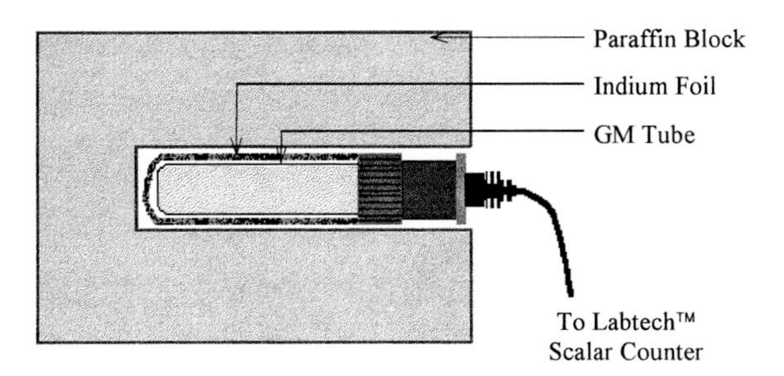

Fig. 3.45 Schematic setup of neutron activation detector using Geiger Müller tube (GM tube) as the detector

of a voltage divider circuit and an emitter follower pre-amp (1:1) for impedance matching. The output pulses of the PMT are counted by a counter through a integrated discriminator.

Alternatively, a Geiger Müller tube (GM tube) can be used as shown in Fig. 3.45 to measure the β^- particles directly. In this case, the indium foil is wrapped around the GM tube [5].

In both setups, the detector is enclosed inside a hydrogenous paraffin block. The function of the paraffin is to thermalize the fusion neutrons which are fast neutrons with energy of 2.45 MeV.

3.6.3 *Time-Resolved Neutron Pulse Measurement*

This is normally done by using a scintillator-photomultiplier system similar to that shown in Fig. 3.46, but without the counting circuit. In this case the output of the photomultiplier is connected to the input of an oscilloscope.

For the measurement of neutrons from high power system, it is necessary to shield the photomultiplier tube and its supporting circuit against electrical noises. This gives rise to some difficulties in the implementation of the scintillation-photomultiplier system. Recently, a system using optical fibre to couple the scintillator to the photomultiplier has been proposed.

This is shown in Fig. 3.46 [5]. In this system, a cylindrical scintillator (NE102) is used for the detection of neutrons. The light produced from the scintillation process is picked up by the fluorescence fibre and its light output is coupled to a glass fibre optic to transmit the light to the photomultiplier tube which is housed inside a screened room together with the oscilloscope. In this way, electrical interference can be minimized.

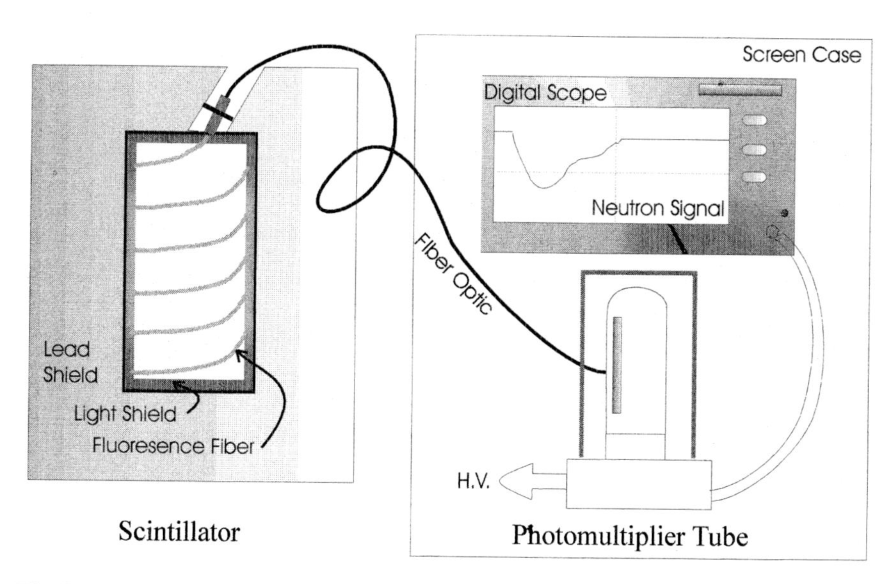

Fig. 3.46 Time resolved neutron detector setup

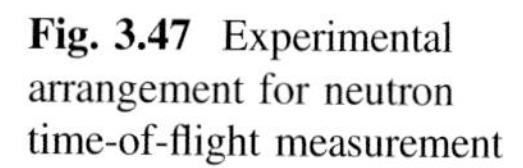

Fig. 3.47 Experimental arrangement for neutron time-of-flight measurement

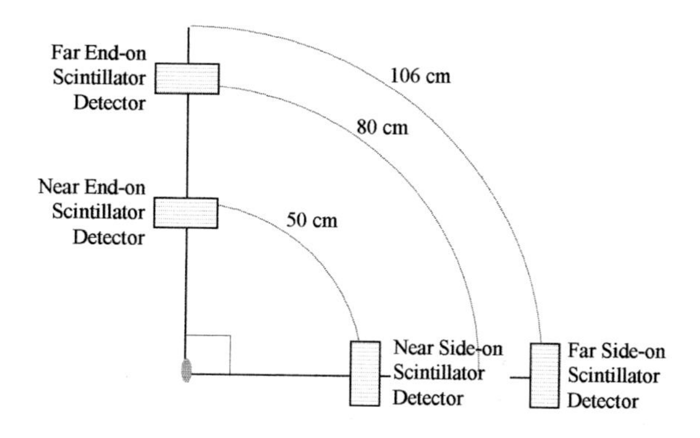

Beside providing information concerning the time evolution of the neutron production, the scintillator-photomultiplier system can also be used to determine the energy of the neutrons. This is done by utilising the time-of-flight technique. The experimental setup is as shown in Fig. 3.47. Two or more channels of scintillator-photomultiplier system are used. With a minimum of 2 channels, placed at different distances from the plasma neutron source, the time taken for neutrons to travel a known distance can be measured from their signals as shown in Fig. 3.48. The speed of the neutrons and hence their energy can then be calculated.

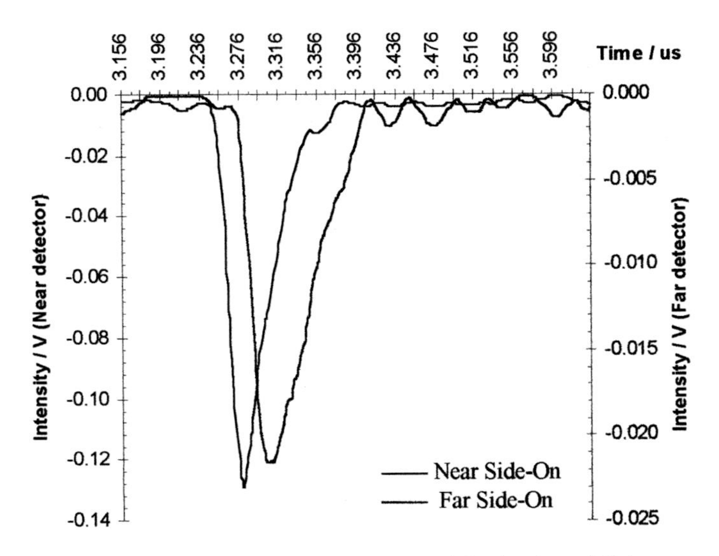

Fig. 3.48 The two well separated neutron pulses obtained for the time-of-flight measurement

References

1. Wong CS (1985) Simple nanosecond capacitive voltage divider. Rev Sci Instruments 56:767–769
2. San Wong C, Choi P, Leong WS, Singh J (2002) Generation of high energy ion beams from a plasma focus modifie for low pressure operation. Jpn J Appl Phys 41:3943–3946
3. Moo SP, Wong CS (1995) Time resolved hard X-ray emission from a small plasma focus. Laser Part Beams 13:129–134
4. Wong CS, Moo SP, Singh J, Choi P, Dumitrescu-Zoita C, Silawatshananai C (1996) Dynamics of X-ray emission from a small plasma focus. Mal J Sci 17B:109–117
5. Yap SL (1998) A study of the temporal and spatial evolution of neutron emission from a plasma focus. MSc Thesis, University of Malaya

General Reference

6. Huddlestone Richard H, Leonard Stanley L (eds) (1965) Plasma diagnostic techniques. Academic Press Inc., New York

Chapter 4
Some Examples of Small Plasma Devices

Abstract In this chapter, the principle of operation, experimental setup as well as some results obtained from research work carried out by the authors and collaborators based on several small plasma devices will be discussed briefly.

Keywords Cost effective plasma devices

4.1 The Electromagnetic Shock Tube [1]

The schematic diagram of an electromagnetic shock tube is shown in Fig. 4.1. The electrodes are of a coaxial configuration. The electrodes are insulated at the back-wall by a quartz cylinder. Discharge is initiated along the surface of the insulator as a surface discharge, forming a current sheet. This current sheet will be pushed away from the insulator surface by the $\boldsymbol{J} \wedge \mathbf{B}$ force until it finally becomes perpendicular to the electrodes and ready to move in the axial direction.

The initial phase of current initiation along the surface of the quartz insulator and lift-off is called the breakdown phase. During the breakdown phase, the gas is mainly heated by the joule heating effect and this is the pre-ionization phase where a sufficiently ionized plasma will be formed prior to the subsequent axial acceleration phase. This is essential since the effective shock heating of the plasma requires it to be sufficiently ionized.

During the axial acceleration phase, the current sheet is driven by the electromagnetic force which is acting as the axial electromagnetic piston. The azimuthal magnetic field B_θ is generated by the coaxial discharge current itself and is given by

$$B_\theta = \frac{\mu I}{2\pi r},$$

which is a function of time and radial position r. This magnetic field is only present behind the current sheet. The magnetic field ahead of the current sheet is zero. It can be seen that its magnitude will be stronger at the surface of the inner electrode as compared to the inner surface of the outer electrode. This gives rise to the slanting

C.S. Wong and R. Mongkolnavin, *Elements of Plasma Technology*,
SpringerBriefs in Applied Sciences and Technology,
DOI 10.1007/978-981-10-0117-8_4

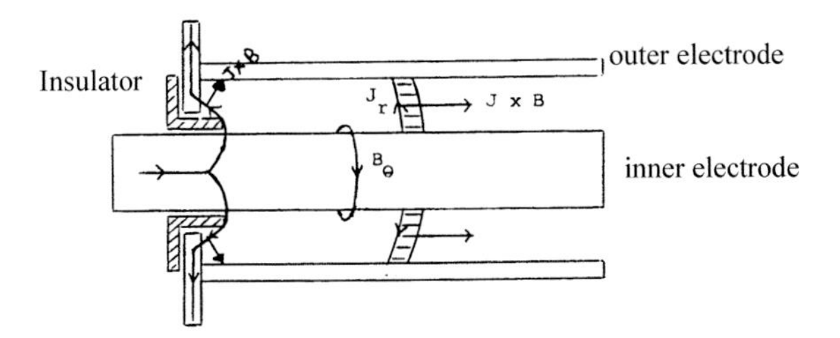

Fig. 4.1 Schematic diagram of an electromagnetic shock tube

structure of the electromagnetic piston as shown in Fig. 4.1. The $\boldsymbol{J} \wedge \boldsymbol{B}$ force in the direction downstream of the tube (in the z-direction) can be expressed as

$$\int_a^b \frac{B_\theta^2}{2\mu} 2\pi r dr,$$

where a is the radius of the inner electrode and b is the radius of the outer electrode. This force will drive the electromagnetic piston to supersonic speed so that a shock heated layer of plasma will be formed. In this way, with discharge current in the region of 100 kA, piston speed of up to more than 10×10^4 m/s may be achieved, which is sufficient to produce a fully ionized hydrogen plasma. There is another component of the $\boldsymbol{J} \wedge \boldsymbol{B}$ force which is acting in the radial direction but no motion is possible due to the present of the solid inner electrode.

4.1.1 Numerical Modeling of the Electromagnetic Shock Tube Dynamics

The dynamics of the electromagnetic piston can be modeled by simplifying the geometry of the shock tube as shown in Fig. 4.2. We assume that (i) the current sheet starts from zero velocity at the position when it first becomes vertical to the electrodes, hence the effective length of the electrodes (z_o) should be measured from the edge of the cylindrical quartz insulator; (ii) the current sheet is not slanting but it is vertical to the electrodes from the surface of the inner electrode to the outer electrode.

The equation of motion for the piston can be written as:

$$\frac{d}{dt}[\rho_1 \pi(b^2 - a^2) z (\frac{dz}{dt})] = \int_a^b \frac{B^2}{2\mu} 2\pi r dr.$$

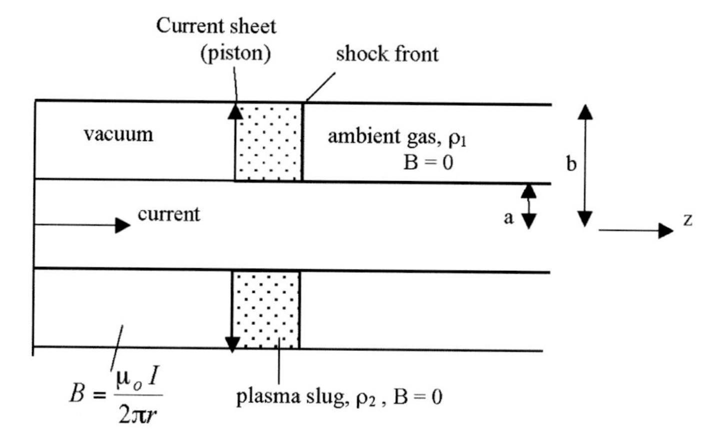

Fig. 4.2 Dynamic model of the electromagnetic shock tube

Substituting the expression for B and re-arranging, we get for the axial acceleration of the current sheet:

$$\frac{d^2z}{dt^2} = \frac{\frac{\mu_o}{4\pi^2\rho_1}\left[\frac{\ell n(b/a)}{b^2 - a^2}\right]I^2 - \left(\frac{dz}{dt}\right)^2}{z}.$$

This differential equation can be solved for the axial current sheet velocity $\left(\frac{dz}{dt}\right)$ if the form of the current as a function of time $I(t)$ is known. Alternatively, the effect of the dynamics of the current sheet on the discharge current can be taken into consideration by writing the coupled circuit equation and solve it simultaneously with the equation of motion. The circuit equation is written by referring to the equivalent circuit of the capacitor discharge system. Assuming the inductive model, where

$$\frac{d}{dt}\left[(L_o + L_p)I\right] \gg I(R_o + R_p).$$

Then the circuit equation is reduced to

$$\frac{d}{dt}\left[(L_o + L_p)I\right] = V_o - \frac{\int_0^t Idt}{C},$$

which can be re-arranged to give the rate of change of current as

$$\frac{dI}{dt} = \frac{V_o - \frac{\int Idt}{C} - \frac{\mu_o}{2\pi}\ell n\left(\frac{b}{a}\right)I\frac{dz}{dt}}{L_o + \frac{\mu_o}{2\pi}\ell n\left(\frac{b}{a}\right)z}.$$

To solve this equation simultaneously with the dynamic equation, it is convenient to first normalize them. This is done by choosing the following normalized parameters:

$$\xi = \frac{z}{z_o}, \quad \tau = \frac{t}{t_o}, \quad \iota = \frac{I}{I_o},$$

where

z_o effective length of the electrodes.

$I_o = V_o\sqrt{\frac{C}{L_o}}$ "short circuit" current.

$t_o = \sqrt{L_o C}$ discharge characteristic time.

The normalized form of the equations are

$$\frac{d^2\xi}{d\tau^2} = \frac{\alpha^2\iota^2 - \left(\frac{d\xi}{d\tau}\right)^2}{\xi},$$

$$\frac{d\iota}{d\tau} = \frac{1 - \int \iota d\tau - \iota\beta\frac{d\xi}{d\tau}}{1+\beta\xi}.$$

α and β are scaling parameters which are given by $\alpha = \frac{t_o}{t_a}$ and $\beta = \frac{L_a}{L_o}$ where

$$t_a = \sqrt{\frac{4\pi^2(b^2 - a^2)\rho_1 z_o^2}{\mu_o \ell n\left(\frac{b}{a}\right) I_o^2}},$$

which is the characteristic dynamic time.

β is the ratio of the maximum tube inductance $L_a = \frac{\mu_o z_o}{2\pi}\ell n\left(\frac{b}{a}\right)$ and the circuit inductance.

The solution can be obtained by using the following boundary conditions:

$$\tau = 0, \quad \xi = 0, \quad \iota = 0, \quad \frac{d\iota}{d\tau} = 1, \quad \int \iota d\tau = 0, \quad \frac{d^2\xi}{d\tau^2} \xrightarrow{\tau \to 0} \frac{\alpha}{\sqrt{2}}.$$

The computation will stop when $\xi = 1$ ($z = z_o$).

An example of the solution obtained is as shown in Fig. 4.3.

These solutions are expected to be not accurate due to the fact that in the actual plasma discharge, the discharge current may be split to form multiple current sheets (current shedding effect) and not all the gas particles are swept by the current sheet (mass shedding effect). In one study [2], it had been shown that the mass swept up by the current sheet to form the plasma can be as little as less than 30 % of all the mass available. These two effects are actually affecting the current sheet dynamics in an opposite manner. While the current shedding effect tends to make the

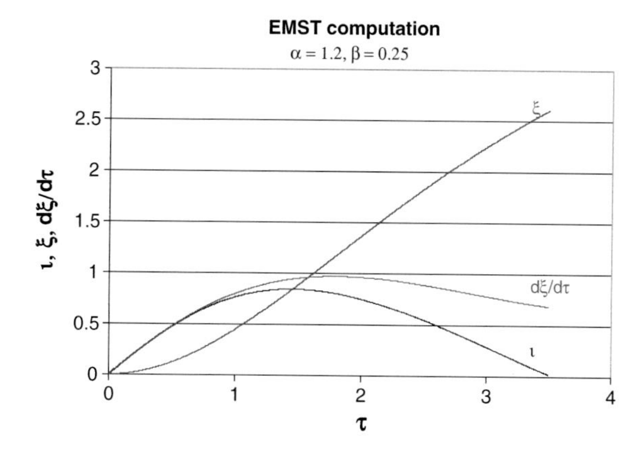

Fig. 4.3 Solutions of the EMST dynamic model

experimentally measured dynamics slower than that predicted by the model, the mass shedding effect tends to make the experimentally measured dynamics faster.

4.1.2 Experimental Measurements of Electromagnetic Shock Tube Dynamics

Since the conditions of the plasma produced by the electromagnetic shock tube are closely related to the dynamics of its current sheet, it is essential in the study of the device to measure the shock speed or the piston speed produced during the discharge. This measurement is commonly done by using the magnetic pickup coil.

The magnetic pickup coil is made of a few turns of copper wire with mm diameter and it "picks up" the localized magnetic flux threading it. Hence it can be used to register the arrival time of the current sheet of an electromagnetic shock tube. This is illustrated in Fig. 4.4.

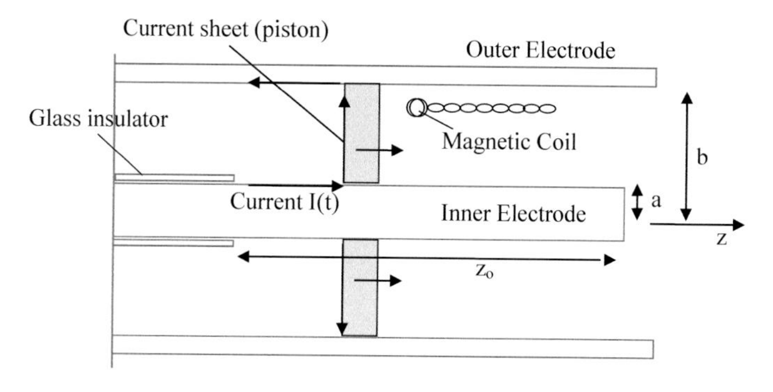

Fig. 4.4 Experimental arrangement for measurement of current sheet dynamics using magnetic probe

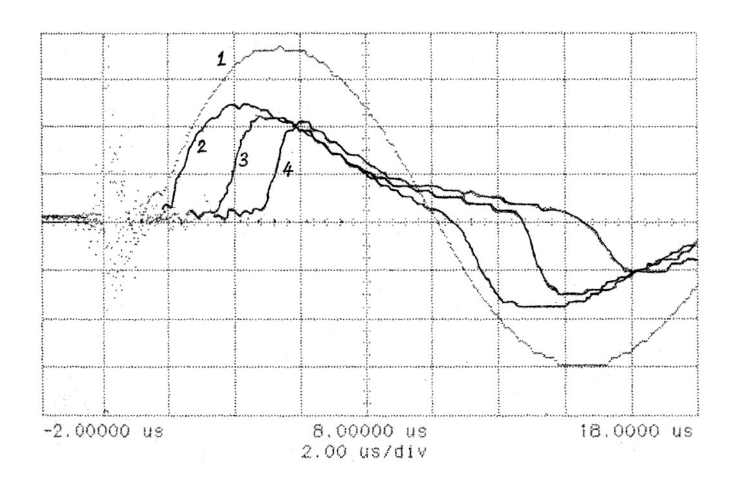

Fig. 4.5 Experimental magnetic probe signals at three positions (waveforms 2, 3 and 4) of the electromagnetic shock tube together with the discharge current (waveform 1)

For the ideal case where the current sheet is assumed to be very thin, the signal registered by the magnetic pickup coil which is corresponding to $\boldsymbol{B}(t)$ is expected to have a sharp rising edge which indicates the arrival of the current sheet. However, the actual current sheet may be diffused and hence the $\boldsymbol{B}$(t) signal registered may look like what is shown in Fig. 4.5 instead.

The signals displayed in Fig. 4.5 are (1) the total discharge current measured by a Rogowski coil; (2) magnetic probe signal at a distance of 2 cm from the back wall of the shock tube; (3) magnetic probe signal at a distance of 6 cm from the back wall of the shock tube; and (4) magnetic probe signal at a distance of 10 cm from the back wall of the shock tube. The times of arrival of the current sheet at various axial positions are easily deduced from these signals.

An alternative method is to measure the rate of change of the magnetic field $\left(\frac{dB}{dt}\right)$ instead of the magnetic field itself. Experimentally, this means removing the *RC* integrator and replaced it by a 50 Ω terminator (or a ×10 resistive voltage divider with a 50 Ω input resistor) instead. The $\left(\frac{dB}{dt}\right)$ signal has significantly larger

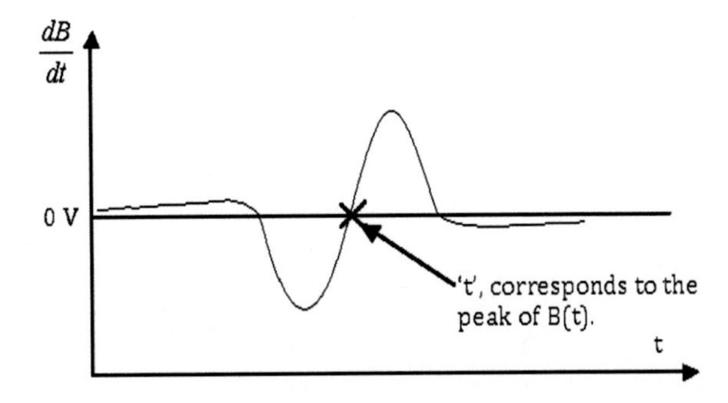

Fig. 4.6 The rate of change of the magnetic probe signal

amplitude as compared to the B signal so it is technically easier to measure. It should be pointed out that in this case the arrival time of the current sheet is taken to be corresponding to the peak of magnetic probe signal, which is the point marked as "X" of the $\left(\frac{dB}{dt}\right)$ signal as shown in Fig. 4.6.

4.2 The Plasma Focus

4.2.1 Introduction

The plasma focus was proposed in the 1960s as a possible device to achieve thermonuclear fusion [2]. In 1962, a research team at the Kurchatov Institute in USSR (Russia) led by Filippov proposed a modification to the linear Z-pinch by providing a inversed pinch phase as a pre-pinch phase. This pre-pinch phase provides some preliminary heating of the plasma before the radial pinching action starts. A similar concept was proposed by Mather and his team in 1964 at the Los Alamos National Laboratory in USA. The Mather design of the plasma focus was originated from the electromagnetic shock tube and in addition to the inverse pinch phase, a second phase of axial acceleration was added as pre-pinch phase. An obvious difference in the two designs is in their geometrical appearances. The Filippov design has a short and large diameter inner electrode while the Mather design has a long and small diameter inner electrode.

Since their respective first reports, the plasma focus device has been studied in numerous laboratories around the world, initially mostly with emphasis on thermonuclear fusion. Currently, the plasma focus is being studied as a possible portable pulsed neutron, radiation (EUV to X-ray) and charged particle (electron and ion) beams source.

4.2.2 Characteristics of the Plasma Focus Discharge [3]

The plasma focus discharge can be described as consisting of three main phases: (1) the inverse pinch (or lift-off) phase, (2) the axial acceleration phase and (3) the radial compression (pinch) phase as indicated in Fig. 4.9b. In the Filippov type plasma focus, the axial phase is practically nonexistence; while in the Mather type, this phase plays a major role in the focus dynamics.

The plasma focus discharge system is basically the same as the electromagnetic shock tube described in Sect. 4.1. The initial phases of the discharge are similar to the electromagnetic shock tube, consisting of the lift-off phase and the axial acceleration phase. However, when the current sheath reaches the truncated end of the electrodes, the current sheath will continue to be pushed axially out of the electrodes thus providing now an additional dimension to be acted upon by the

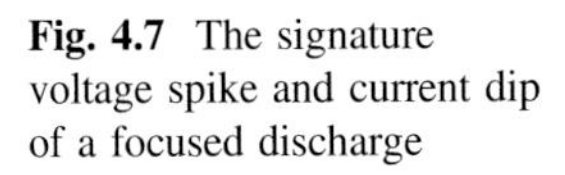

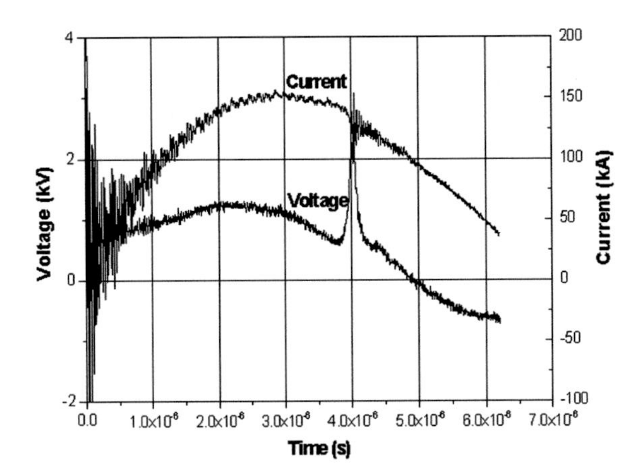

Fig. 4.7 The signature voltage spike and current dip of a focused discharge

electromagnetic force that is radially inwards. This gives rise to what is equivalent to an axially elongating radially compressing Z-pinch. This is the radial compression phase of the plasma focus dynamics. The radial compression occurs very rapidly in a nanoseconds time-scale and the plasma formed at the end of this phase has electron temperature of several keV and electron density of up to 10^{19} cm^{-3}. With such a high temperature and density, the focused plasma is a rich source of radiation emission including X-ray, electron and ion beam and when deuterium is used as the working gas, fusion neutrons will be produced.

The most fundamental parameters of the plasma focus discharge measured experimentally are the discharge current and the voltage across the electrodes. In particular the discharge current is measured by using a Rogowski coil mounted at the back of the focus tubc around the inner electrode, while the voltage drop across the electrodes is measured by using a simple resistive voltage divider. An example of the current and voltage signals for a typical focus discharge is shown in Fig. 4.7. The distinct voltage spike and the current dip are the signatures of the occurrence of focusing action during which the hot dense plasma is produced.

4.2.3 Design Consideration

The first step in the design of the plasma focus system may be considered based on the dynamic model for the axial acceleration phase alone. The main design criterion is to match the rise-time of the discharge current (t_r) to the time of arrival of the current sheet at the end of the inner electrode (t_z) which is actually the time at the end of the axial acceleration phase. t_r is taken to be the quarter period time of the discharge which is given by

$$t_r = \frac{2\pi\sqrt{L_o C}}{4},$$

while t_z is given by

$$t_z \approx 2t_a = 2\sqrt{\frac{4\pi^2(b^2 - a^2)\rho_1 z_o^2}{\mu_o \ell n\left(\frac{b}{a}\right) I_o^2}}.$$

For this two characteristic time to match, we have

$$\frac{64}{\mu_o}\frac{(b^2 - a^2)z_o^2}{\ell n\left(\frac{b}{a}\right)}\frac{\rho_1}{C^2 V_o^2} = 1.$$

Another condition that should be fulfilled is the possible speed that may be achieved. We may decide that the characteristic speed of the current sheet that is necessary to achieve reasonable heating of the plasma is ~10 cm μs^{-1}. By setting

$$U_c = \frac{z_o}{t_a} = \sqrt{\frac{\mu_o \ell n\left(\frac{b}{a}\right) I_o^2}{4\pi^2(b^2 - a^2)\rho_1}} = 10^5\ (\mathrm{ms}^{-1}),$$

the matching condition becomes

$$z_o = \frac{\pi}{4} \times 10^5 \sqrt{L_o C}.$$

We may start with a given capacitor and an estimate of system inductance including that of the capacitor itself, then the length of the electrode is fixed by the expression above. For example, if a capacitor of capacitance 30 μF is available. Lets say we estimate the inductance of the system can be kept at not more than 110 nH, then an electrode length of 0.14 m may be appropriate.

With C and z_o fixed, ρ_1, V_o, b and a can then be matched according to

$$\frac{64}{\mu_o}\frac{(b^2 - a^2)z_o^2}{\ell n\left(\frac{b}{a}\right)}\frac{\rho_1}{C^2 V_o^2} = 1.$$

A chart of these parameters that may match this condition can be drawn up with the constraint that $a < b \ll z_o$. It is probably logical to fix V_o first according to the capacitor available, then the operating gas pressure (ρ_1). This allows us to seek for a suitable combination of a and b. This procedure can be repeated for other possible values of V_o and ρ_1.

Finally, experimental fine tuning may still be necessary to arrive at a set of parameters for optimum plasma focus operation.

4.2.4 X-ray Emission from the Plasma Focus Discharge

During the plasma focus discharge, X-rays are emitted due to two possible mechanisms [4–6]. The first is the hot dense plasma produced at the end of the radial compression phase. With conditions of electron temperature $T_e \sim$ keV and density of $n_e \sim 10^{19}$ cm^{-3}, the plasma is expected to emit photons with energy in the X-ray region. The peak of the continuum (Bremsstrahlung and recombination) is expected to be several Å while the line radiations are due to inner shell transition (K_α, K_β of highly ionized species) which will be in the X-ray region. The second mechanism is due to the interaction of high energy electron beam with the inner electrode surface. Due to the development of MHD instabilities soon after the end of the radial compression phase, high localized electric field may be induced and high energy ion beam and electron beam may be produced. The electron beam will bombard at the surface of the inner electrode giving rise to the generation of hard X-ray, which is predominantly the K_α line radiation of the electrode material (in most cases copper). The line radiation may also be due to contamination of the focused plasma contributed by the electron beam sputtered copper vapor from the anode [7].

In any typical plasma focus discharge, both these two types of X-ray emission may be observed as shown in Fig. 4.8. In this discharge, X-ray pulse is observed both coincide with the voltage spike as well as at time beyond 50 ns after the voltage spike.

The X-ray emitting region of the plasma focus is observed to have a complicated structure. Normally the tip of the inner electrode can be seen in the X-ray image obtained experimentally. The X-ray image of the plasma column is observed to be above the inner electrode, often with spot-like structures superimposed on it [8, 9]. An example of the X-ray images of the plasma focus is shown in Fig. 4.9.

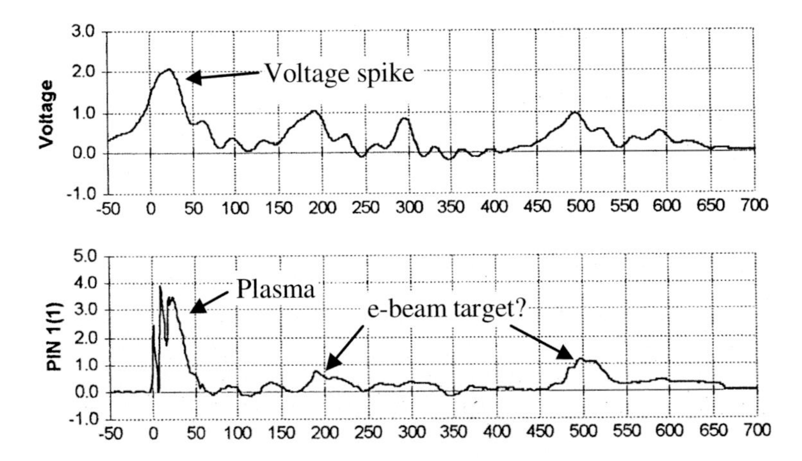

Fig. 4.8 Evidence of two types of X-ray emission from a focused plasma

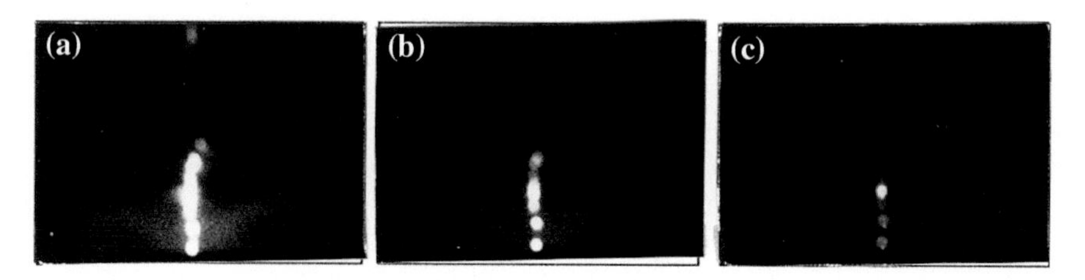

Fig. 4.9 Images of spot-like plasma within the focused plasma

This set of X-ray images are obtained for the same plasma focus discharge but with different absorption filters: (a) 48 μm aluminized Mylar; (b) 48 μm aluminized Mylar + 10 μm Al; and (c) 48 μm aluminized Mylar + 20 μm Al. These filters allow regions of different X-ray hardness to be revealed. It is evident that within the plasma column in (a), minute spots emitting harder X-ray are present as shown in (b) and (c). The shape of these X-ray spots is round indicating that their size is probably smaller than the pinhole used in the pinhole camera setup.

A possible application of the plasma focus X-ray is in X-ray lithography. The feasibility of such application has been tested in various laboratories. High Z gases such as Xenon and Krypton are often used for the operation of the plasma focus as a pulsed X-ray source [10, 11].

4.2.5 Neutron Emission

When deuterium is used as the operating gas, the plasma focus has been shown to be capable of achieving fusion condition at the end of the radial compression phase. This gives rise to the generation of neutrons. However, there are evidences that the neutrons emitted from the focused plasma is not totally of thermonuclear origin. This is in fact expected since the final temperature achieved in the plasma focus is only of the order of few keV. Any fusion reaction must be due to deuterons with kinetic energy at the high energy tail of the energy spectrum of the plasma. A large percentage of the neutrons measured, probably more than 60 % in the case of a small plasma focus, are actually produced by the beam target effect. High energy ion beam, in this case deuteron beam, is generated due to localized electric field induced by instabilities.

One of the evidences of non-thermonuclear neutron emission is that the energy of the neutron is 2.8 MeV instead of 2.45 MeV as measured by the neutron time-of-flight technique as described in Sect. 3.6.3 earlier.

Secondly, the neutron emission angular distribution is found to be anisotropic, with the ratio of neutron yield measured in the end-on direction (0°) to that in the side-on direction (90°) to be greater than 1. For example, in a series of measurements, the neutron yields measured end-on and side-on simultaneously give a ratio of 1.5 [12].

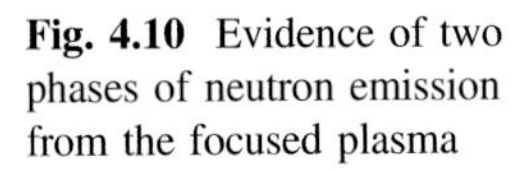

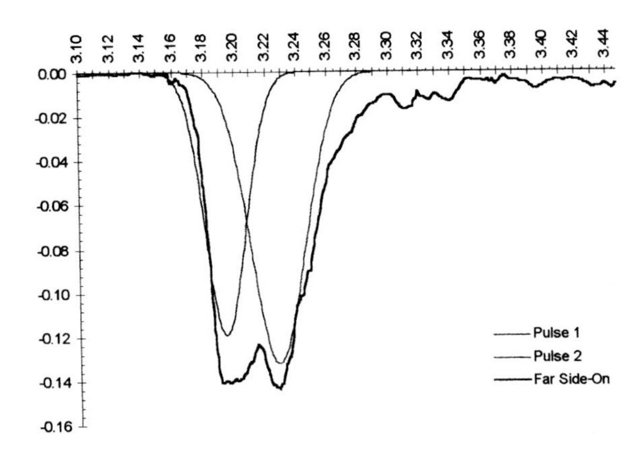

Fig. 4.10 Evidence of two phases of neutron emission from the focused plasma

Thirdly, from the analysis of the time-resolved neutron pulse obtained experimentally, it can be shown by using the Gaussian curve fitting technique that the neutron pulse can be resolved into two Gaussian pulses (Fig. 4.10) indicating the possibility of 2 neutron components. The first pulse which is more aligned to the voltage spike is believed to be thermonuclear in nature while the second pulse is interpreted as produced by the deuteron beam target mechanism [13].

4.2.6 Particle Beam Emission

The existence of energetic electron beam in the plasma focus is believed to be closely co-related to phenomena that give rise to localized high electric field, such as the rapid compression during the radial pinch phase and instability. Both mechanisms are found to be present and indeed two periods of electron beam generation have been observed. The electron beam observed during the first period is believed to be associated with the first maximum compression and is measured to be having energy as high as 180 keV. The second period electron beam, which is tied to the $m = 0$ instability, has a much lower energy [14, 15].

Together with the generation of electron beam, the same localized induced electric field is also responsible for the generation of energetic ion beam [16, 17]. Ion beams with energies of up to MeV have been reported in the literature. The average nitrogen ion beam measured is about 180 keV, which seems to be consistent with the averaged energy of the electron beam measured. The applications of the plasma focus generated ion beam to modify the nano-scaled structure of various material surfaces have been widely reported recently [18–20].

4.3 The Vacuum Spark

The vacuum spark is known to be a source of intense pulsed X-ray as well as a source of highly-stripped ions such as Fe-XXVI, Ni-XXVII and Mo-XIII-XVIII. In view of the close resemblance of its X-ray spectrum with that of solar flares, the vacuum spark is also being used to simulate the solar flare phenomenon in the laboratory.

The vacuum spark has a relatively simple configuration as compared to the plasma focus. It consists of a pair of electrodes kept at a distance of less than 1 cm apart and placed in vacuum (typically $P < 10^{-3}$ mbar is sufficient). The electrodes are connected directly across a capacitor so no switch is used. When the capacitor is charged to a high voltage V (typically from 10 to 30 kV), breakdown across the inter-electrode gap is prohibited by the high vacuum since the value of the product of pressure and inter-electrode spacing, pd, is on the left hand side of the Paschen curve minimum. Discharge is initiated by injecting a puff of vapor of the anode material (partially ionized) by one of the following methods: (1) by using a sliding spark between the cathode and a third electrode to produce some electrons which will be accelerated by the electric field to bombard at the anode to produce the anode vapor [21]; (2) by focusing a high power pulsed laser beam to vaporize the anode material directly [22]; and (3) by utilizing the Transient Hollow Cathode electron beam to vaporize the anode material [23, 24].

The vacuum spark plasma has been measured to achieve temperature of as high as 8 keV and density of 10^{20}–10^{21} cm^{-3}. The structure of the plasma formed may consist of plasma cloud and minute spots with a source size of 100–200 μm, an example of which is shown in Fig. 4.11. In this image, the tip of the anode is also seen to be emitting X-ray, probably due to bombardment by electrons.

The X-ray emission spectrum of the vacuum spark plasma is found to be consisting of a mixture of continuum and line radiations. Its spectral contents can be classified into 3 categories, namely plasma emission dominated (continuum radiation dominated), electron beam—target emission dominated (line radiation dominated) or mixture of both types of emission. For the plasma emission dominated discharges, the electron temperatures are estimated to be in the range of 3–10 keV. For discharges where the X-ray spectrum is dominated by line radiation, the plasma emission is relatively weak and the X-ray photons are predominantly produced by

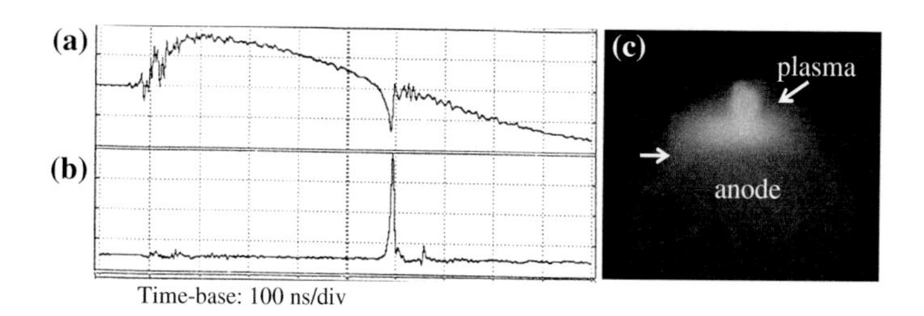

Fig. 4.11 **a** *dI/dt* signal, **b** X-ray pulse, **c** X-ray pinhole image of a typical vacuum spark plasma

electron beam bombardment of the anode which gives rise to strong K_{α} emission of the anode material. The intense energetic electron beam may be produced by either the pre-breakdown transient hollow cathode discharge which is utilized to initiate the main discharge, or the strong m = 0 instability at the final stage of the pinching effect of the plasma.

4.4 Scaled-Down Operation of Vacuum Spark—Flash X-ray Tube

In view of the observation of electron beam target line radiation emission during the pre-breakdown phase of the vacuum spark induced by the transient hollow cathode effect, the vacuum spark system can be scaled down in terms of electrical input energy to operate as a flash X-ray tube [25–27]. The schematic diagram of the flash X-ray tube is illustrated in Fig. 4.12.

The operation of the flash X-ray tube is exactly the same as the vacuum spark but the main discharge current is low so the plasma produced will not be hot enough to produce X-ray. As can be seen from Fig. 4.12, the discharge is powered by eight ceramic capacitors with capacitance of 2.7 nF each, making up a total capacitance of 21.6 nF as compared to the 1.85 μF capacitor used in the high energy version. When discharged at a voltage of 20 kV, the flash X-ray tube input electrical energy is only 4.32 J as compared to 370 J in the case of the vacuum spark. However, with

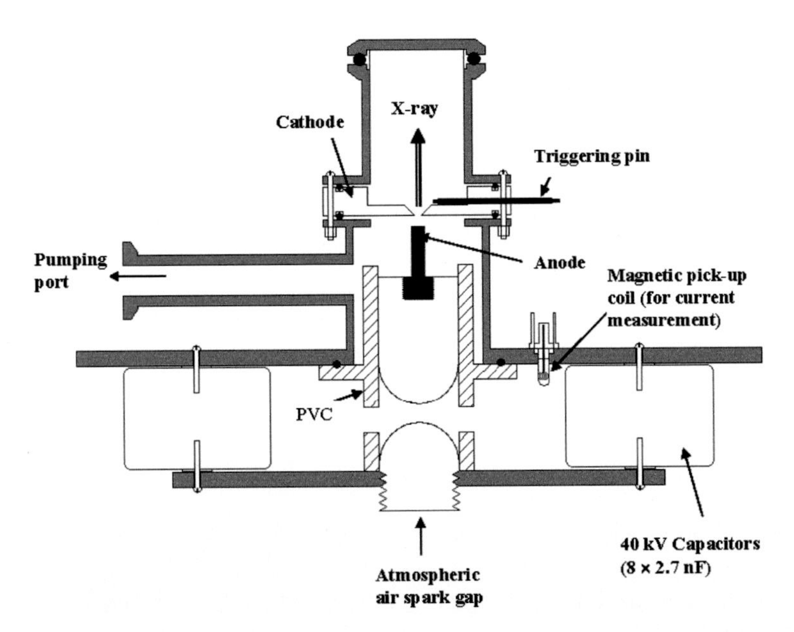

Fig. 4.12 Schematics of a flash X-ray tube. Reprinted from Ref. [26], Copyright (2007), with permission from Cambridge University Press

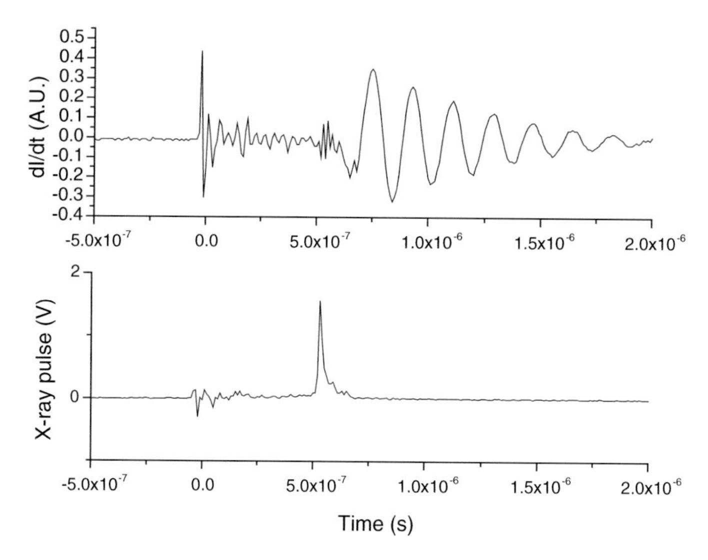

Fig. 4.13 Example of the discharge current waveform and the X-ray pulse of a flash X-ray tube discharge. Reprinted from Ref. [26], Copyright (2007), with permission from Cambridge University Press

the same capacitor discharge voltage applied across the electrodes, the intensity of the electron beam target X-ray emission due to bombardment of the transient hollow cathode electron beam is expected to be the same in the two cases. An example of the X-ray pulse produced by the flash X-ray tube is as shown in Fig. 4.13.

Some of the applications of the flash X-ray tube as X-ray source for radiography of small biological sample [26] as well as for testing of thermo-luminescence response of fiber dosimeter has been reported recently [28, 29].

The X-ray emission spectrum of the flash X-ray tube can be easily tuned by changing the anode material. Figure 4.14 shows the X-ray spectra of (from top to bottom) Al, Ti, Cu and W [26].

4.5 The Pulsed Capillary Discharge

The pulsed capillary discharge has been shown to be a copious source of EUV and soft X-ray. Recently, due to the successful demonstration of amplification of the 3 s $^1P_1$–3p $^1S_0$ line of Ne-like Ar at $\lambda = 46.9$ nm in a capillary discharge, this device has attracted much research interests from various researchers.

The pulsed capillary discharge is basically a Z-pinch discharge restricted to occur within the small diameter capillary channel. An example of a capillary discharge powered by capacitor discharge [30] is shown in Fig. 4.15a. The discharge of this system is powered by six 30 kV, 3.6 nF doorknob capacitors connected in

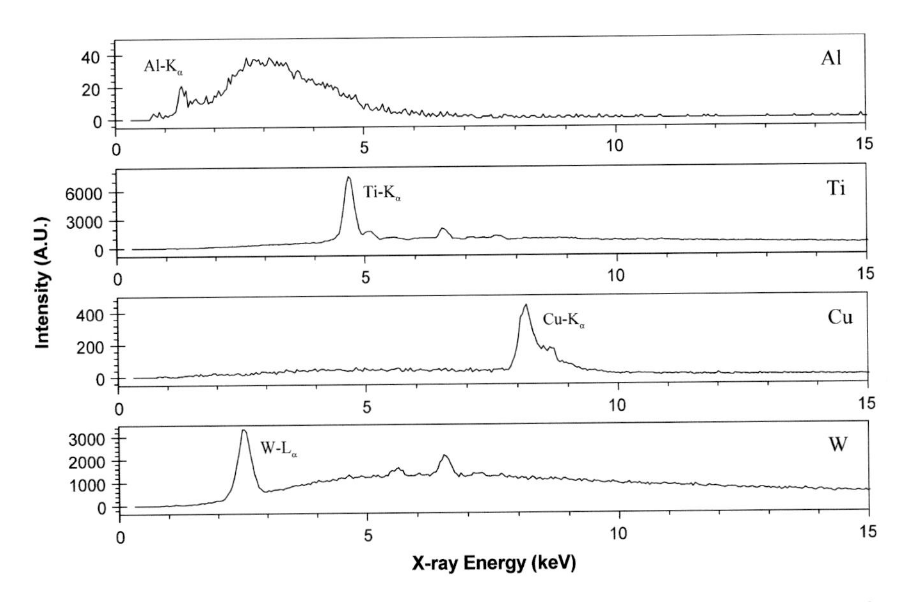

Fig. 4.14 Examples of X-ray emission spectra of the flash X-ray tube. Reprinted from Ref. [26], Copyright (2007), with permission from Cambridge University Press

parallel to give a maximum discharge input energy of 9.7 J. The capillary is made of quartz tube with inner diameter of 1 mm and a length of 10 mm. The system is pumped to a pressure of 10^{-5} mbar. To perform a discharge, the capacitor is first charged up to the desired high voltage. A high voltage pulse is then applied to the triggering pin to initiate the transient hollow cathode discharge in the hollow cathode region which subsequently leads to the main discharge through the capillary.

The discharge current waveform together with radiation emission pulses from a typical 24 kV discharge of this capillary discharge system are shown in Fig. 4.15b. The discharge current is measured by the magnetic pick up coil (the magnetic probe), while the X-ray pulse is measured by a PIN diode and the EUV pulse is measured by a silicon photodiode sensitive to the EUV range. The peak current reached is about 8 kA.

From Fig. 4.15b, it can be seen that the discharge can be divided into two major phases: the initiation phase and the main discharge phase. During the initiation phase, before the main discharge set in, the discharge current is started with a sharp pulse following by a slow rising edge. Initially the 24 kV high voltage is divided between the capillary and the external spark gap. Before the occurrence of the full discharge, due to the voltage held across the cathode and the anode, electrons from the triggering pulse in the hollow cathode region may be attracted to flow across the capillary to hit at the anode. This produces the low level X-ray emission observed in the X-ray pulse. The small sharp current pulse corresponds to the external spark gap firing which transfers the full voltage across the capillary. This further enhance the

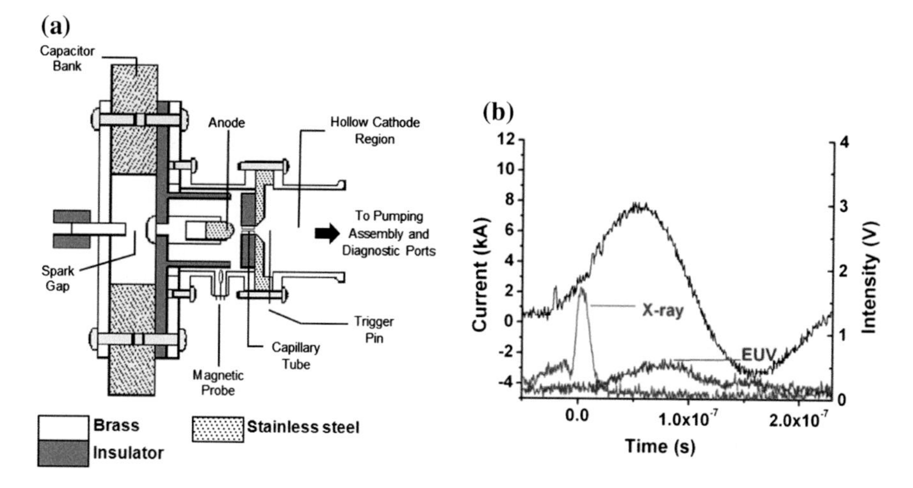

Fig. 4.15 **a** Schematics of the pulsed capillary discharge device. **b** The discharge current, pre-breakdown electron beam induced X-ray pulse and EUV pulse emitted from the plasma. Figure 4.15a is reprinted from Ref. [30], Copyright (2014), with permission from Elsevier

electron beam activity and eventually leads to avalanche of electrons forming a beam which bombards at the anode to give a large X-ray signal. Breakdown of the main discharge gap across the capillary will then occur. The formation of the capillary discharge plasma is accompanied by the emission of 30 mJ of radiation energy in the wavelength range of 11–18 nm of the EUV spectrum. The electron temperature of this pulsed capillary discharge plasma is estimated to be about 10 eV.

4.6 The 50 Hz Alternating Current (A.C.) Glow Discharge System

There are two major cost factors in any plasma system. First is the requirement of low or ultra low operating pressure. The creation of high vacuum means large budget is needed to purchase high vacuum equipments, which consist of the pumping system as well as the pressure gauges, both of which are expensive. It is for this reason that the development of atmospheric pressure discharges has attracted much research efforts in recent years. The second factor is the power source. Comparing with the RF, microwave or even the DC power sources, the relatively cheap power source is the 50 Hz AC power source which can be obtained directly from the domestic power source via a step-up transformer. Another major advantage of using the 50 Hz AC power source is that no impedance matching network is needed, which is an important cost reducing factor. For the measurement of the discharge current, a simple milli-ammeter or a multi-meter can be used. Since

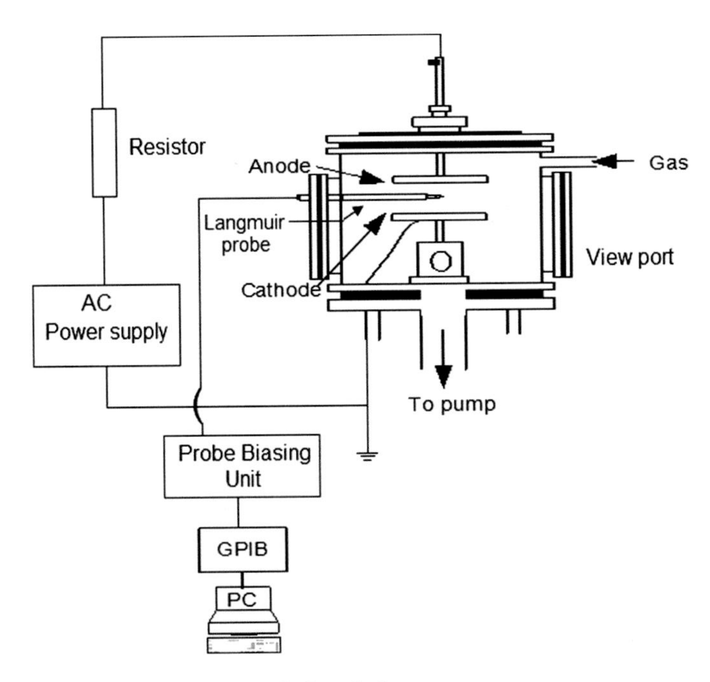

Fig. 4.16 Schematics of the 50 Hz AC glow discharge system

the discharge voltage required is normally in the kV region, a voltmeter coupled with a resistive voltage divider may be used.

An example of the 50 Hz glow discharge system is shown schematically in Fig. 4.16. The system consists of the following components: The discharge chamber, a rotary pump, a step-up transformer and the gas flow control. The electrodes are two circular stainless steel plates. The operation of the system is straight forward. After breakdown, the discharge current is controlled by using the current limiting resistor R to be within the range of glow discharge, normally in the mA region.

The electron temperature and density of the plasma produced can be measured by using the Langmuir probe as indicated in Fig. 4.16. In a series of experiments, argon glow discharge plasma was produced by using a discharge voltage of 16 kV peak to peak, while the working pressure was varied from 0.3 to 0.6 mbar. Various sets of I–V characteristics were obtained with the Langmuir probe placed at the center of the plasma column. The experiment was done with an inter-electrode distance of 2 cm. An example of the I–V characteristics obtained is displayed in Fig. 4.17.

The variations of electron density and electron temperature with pressure for inter-electrode distance 2 cm are shown in Fig. 4.18.

It is seen that there is an increase of electron density and decrease of electron temperature with increase in pressure from 0.3 to 0.6 mbar. The maximum electron density and electron temperature are found to be 3.58×10^{16} m^{-3} and 5.98 eV,

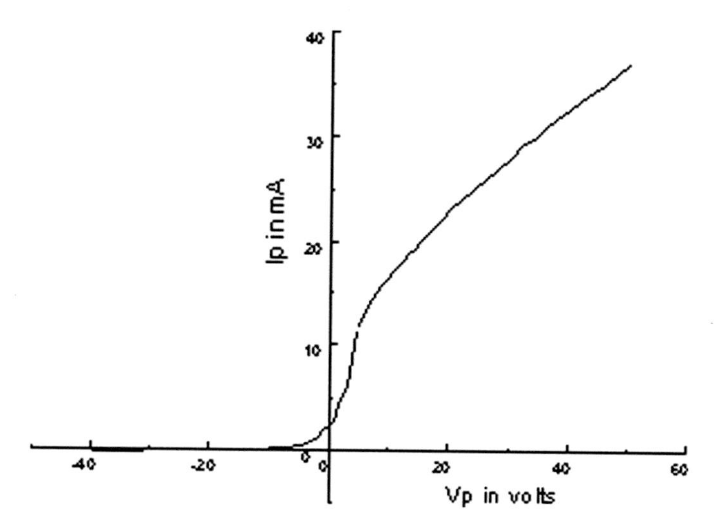

Fig. 4.17 An example of typical I–V characteristics of Langmuir probe

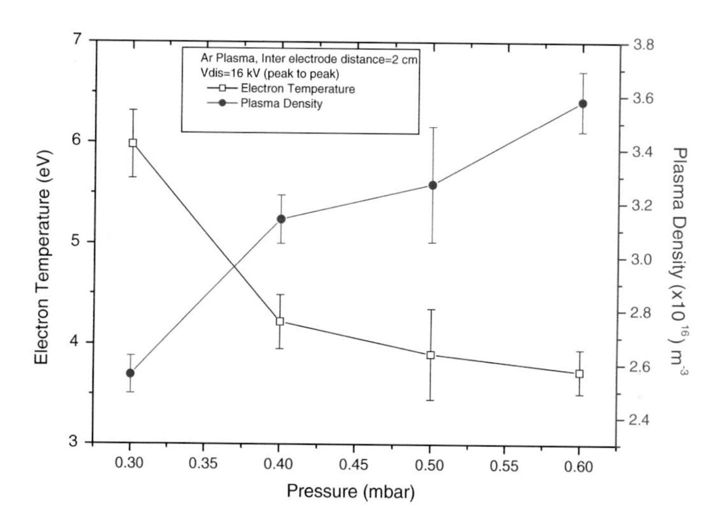

Fig. 4.18 Variation of electron temperature and density with pressure

respectively. The increase of electron density with pressure is believed to be due to more ionization. Since increase in pressure increases the neutral particle density, so the probability of energy loss of the plasma electrons due to the collision with neutral particle increases. So the energy of the electrons decreases i.e., electron temperature decreases with increase in pressure. It is noticed that plasma density increases in case of 2 cm inter-electrode distance compared to that of 1.5 cm for a particular pressure. On the other hand, electron temperature decreases for a particular pressure in case of 2 cm inter-electrode distance compared to that of 1.5 cm.

It is believed that when the inter-electrode distance increases, the electrons undergo more collisions resulting in an increase of plasma density. Because of more collisions, electrons loss energy and that is why electron temperature decreases with an increase of inter-electrode distance.

The 50 Hz glow discharge plasma has been used successfully to treat bio-materials such as gelatin and Thai silk fibroin to improve the wettability of their surfaces [31–35].

4.7 Atmospheric Pressure Dielectric Barrier Discharge

The requirement of a low pressure ambient gas environment for the generation of plasma is one of the important cost factor in the application of plasma technology in industry. In view of this consideration, there is much research interest to investigate the feasibility of using atmospheric pressure plasmas for various processes instead of low pressure plasmas. One type of device that fit well into this criterion is the dielectric barrier discharge which is being studied in many small laboratories worldwide.

The dielectric barrier discharge, or DBD in short, can be constructed in two types of configuration, either with parallel plate electrodes or coaxial electrodes as illustrated in Fig. 4.19.

A common feature of these two setups is the present of a dielectric between the electrodes so that when an oscillating electric potential is applied across the electrodes, intermittent electric current pulses may occur between the surface of the dielectric barrier and one of the electrodes. These current pulses are normally occurring in burst at both the positive and negative half cycles of the alternating voltage as shown in Fig. 4.20 [36, 37]. Each of the current pulse may have a pulse width of less than 200 ns as shown in Fig. 4.21 [38]. Similar discharge characteristics are also observed with the coaxial DBD.

The cylindrical coaxial DBD has been found to be particularly suitable for applications as chemical reactor for gases. One example of using the coaxial DBD as a chemical reactor and it has been tested to be effective for converting oxygen to ozone [39] and also for dissociating nitric oxide [40, 41].

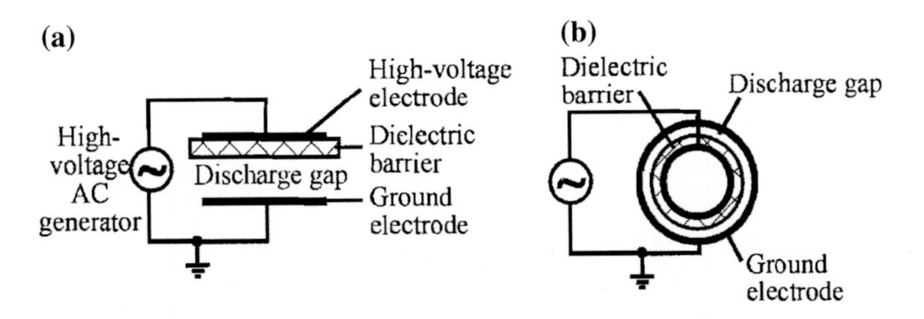

Fig. 4.19 Two possible configurations of dielectric barrier discharge

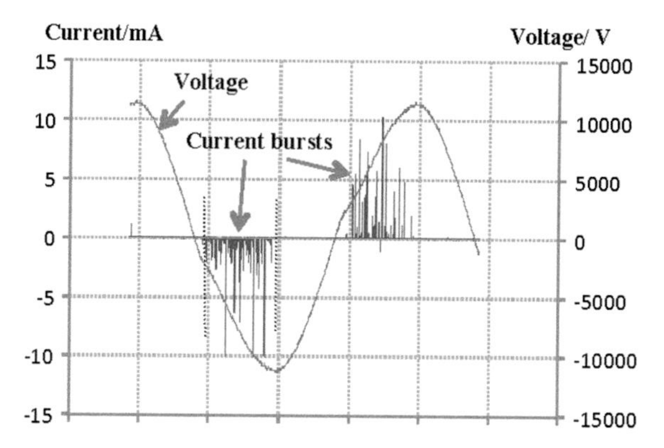

Fig. 4.20 Two bursts of current pulses and the corresponding sinusoidal voltage waveform of a 50 Hz dielectric barrier discharge

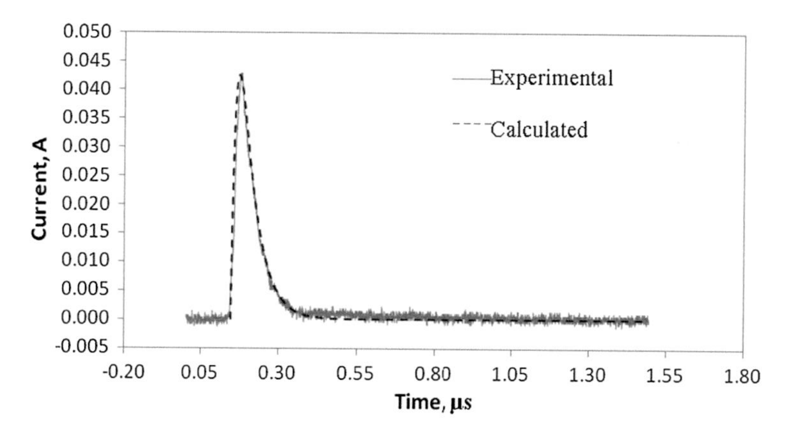

Fig. 4.21 Single current pulse with ~200 ns pulse width

For the application of planar DBD such as treatment of material surface, a more uniform planar glow discharge type plasma is required. For such an application it may be necessary to use a voltage source of kHz frequency [42].

4.8 Wire Explosion System for Nano Powder Fabrication

The wire explosion process is a simple technique that can be used effectively to synthesize nano particles of pure metals or their composites. The pulsed discharge system is similar to those used for the vacuum spark as shown schematically in Fig. 4.22. In this particular system, the discharge is powered by a high voltage capacitor with rating of capacitance 1.85 μF and discharge voltage of up to 50 kV.

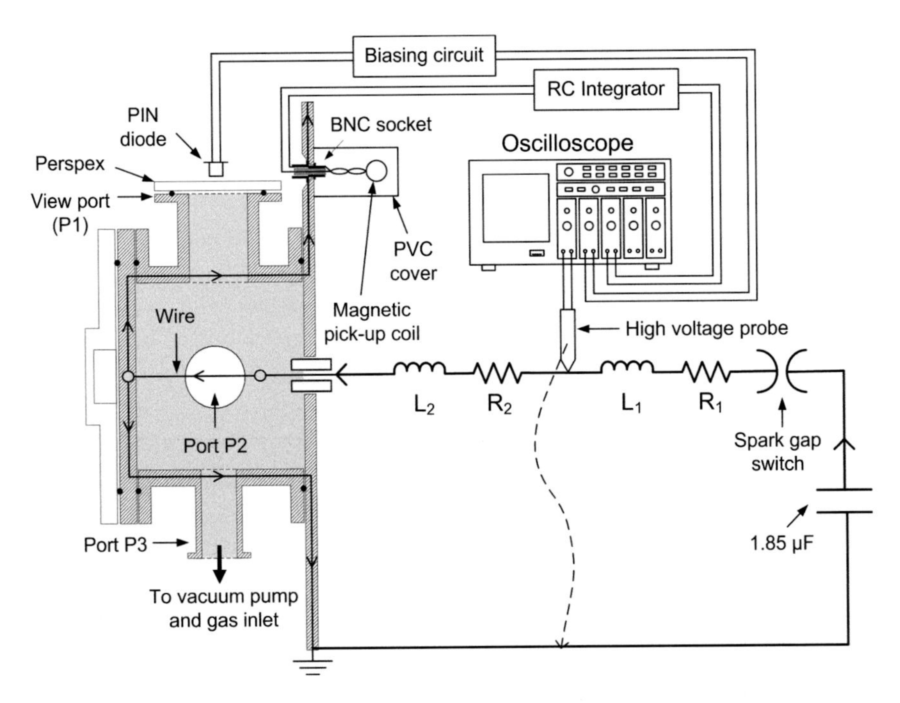

Fig. 4.22 Schematics of a wire explosion system

However, for the purpose of the wire explosion process, a nominal discharge voltage of 15 kV is sufficient. The chamber has a diameter of 9.4 cm and a height of 10.3 cm. Magnetic pick-up coil and high-voltage probe are used to register the current and voltage signals respectively during the wire explosion, while a PIN diode (BPX65) can be used to record the light (visible to ultraviolet) emitted during the wire explosion and the subsequent electrical discharge through the vapor. The energy deposited into the wire can be calculated from the measured current and voltage signals. The measured current density through the wire during the explosion of the wire is found to be of the order of 106 A cm^{-2}. This conduction continues until the vapors are optically thick and only radiation emitted from the surface will be detected by the PIN diode. At this stage, a column of ionized plasma is formed. Subsequently, the plasma begins to expand in the background of the ambient gas due to the enormous difference in the temperature and pressure between the plasma and the ambient gas. The expanded plasma particles are rapidly cooled down during this process of expansion and leading to the formation of a supersaturated vapor, which will undergo a homogeneous nucleation of nano particles. The ambient gas type as well as its pressure have been found to have great influence on the nano particle formation [43–47].

References

1. Lee S (1985) Experiment 1: electromagnetic shock tube. In: Lee S, Tan BC, Wong CS, Chew AC (eds) Laser and plasma technology. World Scientific Publishing Co. Pvt. Ltd, Singapore
2. Bernard A, Bruzzone H, Choi P, Chuaqui H, Gribkov V, Herrera J, Hirano K, Krejci A, Lee S, Luo C, Mezzetti F, Sadowski M, Schmidt H, Ware K, Wong CS, Zoita V (1998) Scientific status of plasma focus research. J. Moscow Phys. Soc. 8:93–170
3. Lee S (1985) Experiment II: plasma focus experiment & technology of plasma focus. In: Lee S, Tan BC, Wong CS, Chew AC (eds) Laser and plasma technology. World Scientific Publishing Co. Pvt. Ltd, Singapore
4. Choi P, Wong CS, Herold H (1989) Studies of the spatial and temporal evolution of a dence plasma focus in the X-ray Region. Laser Part Beams 7:763–772
5. Favre M, Lee S, Moo SP, Wong CS (1992) X-ray emission in a small plasma focus operating with H_2-Ar mixtures. Plasma Sources Sci Technol 1:122–125
6. Wong CS, Moo SP, Singh J, Choi P, Dumitrescu-Zoita C, Silawatshananai C (1996) Dynamics of X-ray emission from a small plasma focus. Mal J Sci 17B:109–117
7. Al-Hawat Sh, Akel M, Wong CS (2011) X-ray emission from argon plasma focus contaminated with copper impurities in AECS PF-2 using five channel diode spectrometer. J Fusion Energy 30:503–508
8. Ng CM, Moo SP, Wong CS (1998) Variation of soft X-ray emission with gas pressure in a plasma focus. IEEE Trans Plasma Sci 26:1146–1153
9. Kulkoulprakarn Titisak, Ngamrungroj Dusit, Kamsing Pirud, Wong Chiow San, Mongkolnavin Rattachat (2007) X-ray source structures of a small plasma focus device. J Sci Res Chula Univ 32:55–60
10. Mohammadi MA, Verma R, Sobhanian S, Wong CS, Lee S, Springham SV, Tan TL, Lee P, Rawat RS (2007) Neon soft X-ray emission studies from UNU-ICTP plasma focus operated with longer than optimal anode length. Plasma Sources Sci Technol 16:785–790
11. Mohammadi MA, Sobhanian S, Wong CS, Lee S, Lee P, Rawat RS (2009) The effect of anode shape on neon soft X-ray emissions and current sheath configuration in plasma focus device. J Phys D Appl Phys 42:045203
12. Yap SL, Wong CS (2007) Development of a 3.3 kJ plasma focus as pulsed neutron source. J Sci Technol Tropics 3:123–127
13. Yap SL, Wong CS, Choi P, Dumitrescu C, Moo SP (2005) Observation of two phases of neutron emission in a low energy plasma focus. Jpn J Appl Phys 44:8125–8132
14. Choi P, Deeney C, Wong CS (1988) Absolute timing of relativistic electron beam in a plasma focus. Phys Lett A 128:80–83
15. Choi P, Deeney C, Herold H, Wong CS (1990) Characterization of self-generated intense electron beams in a plasma focus. Laser Part Beams 8:469–476
16. Lee CH, Ngamrungroj D, Wong CS, Mongkolnavin R, Low YK, Singh J, Yap SL (2005) Correlation between the current sheath dynamics in the axial acceleration phase of the plasma focus and its ion beam generation. J Sci Technol Tropics 1:51–54
17. Lim LK, Yap SL, Wong CS, Zakaullah M (2013) Deuteron beam source based on Mather type plasma focus. J Fusion Energy 32:287–292
18. Ngoi SK, Yap SL, Goh BT, Ritikos R, Rahman SA, Wong CS (2012) Formation of nano-crystalline phase in hydrogenated amorphous silicon thin film by plasma focus ion beam irradiation. J Fusion Energy 31(1):96–103
19. Goh BT, Ngoi SK, Yap SL, Wong CS, Dee CF, Rahman SA (2013) Structural and optical properties of the Nc–Si: H thin films irradiated by high energetic ion beams. J Non-Cryst Solids 363:13–19
20. Goh BT, Ngoi SK, Yap SL, Wong CS, Rahman SA (2013) Effect of energetic ion beam irradiation on structural and optical properties of A–Si: H thin films. Thin Solid Films 529:159–163

21. Lee S, Conrad H (1976) Measurements of neutrons and X-rays from a vacuum spark. Phys Lett 57A:233–236
22. Wong CS, Lee S (1984) Vacuum spark as reproducible X-ray source. Rev Sci Instrum 55:1125–1128
23. Wong CS, Ong CX, Lee S, Choi P (1992) Observation on enhanced pre-breakdown electron beams in a vacuum spark with a hollow cathode configuration. IEEE Trans Plasma Sci 20:405–409
24. Wong CS, Ong CX, Moo SP, Choi P (1995) Characteristics of a vacuum spark triggered by the transient hollow cathode discharge electron beam. IEEE Tran Plasma Sci 23:265–269
25. Wong CS, Lee S, Ong CX, Chin OH (1989) A compact low voltage flash X-ray tube. Jpn J Appl Phys 28:1264–1267
26. Wong CS, Woo HJ, Yap SL (2007) A low energy tunable pulsed X-ray source based on the pseudospark electron beam. Laser Part Beams 25:497–502
27. Wong CS, Singh J (2012) Malaysian patent: pulsed plasma X-ray source, MY-145318-A
28. Amin YM, Mahat RH, Donald D, Shankar P, Wong CS (1998) Measurement of X-ray exposure from a flash X-ray tube using TLD 100 and TLD 200. Radiat Phys Chem 51:479
29. Bradley DA, Wong CS, Ng KH (2000) Evaluating the quality of images produced by soft X-ray units. Appl Radiat Isot 53:691–697
30. Chan LS, Tan D, Saboohi S, Yap SL, Wong CS (2014) Operation of an electron beam initiated metallic plasma capillary discharge. Vacuum 103:38–42
31. Wong CS, Lem SP, Goh BT, Wong CW (2009) Electroless plating of copper on polyimide film modified by 50 Hz plasma graft polymerization with 1-Vinylimidazole. Jpn J Appl Phys 48:036501
32. Prasertsung I, Mongkolnavin R, Damrongsakkul S, Wong CS (2010) Surface modification of dehydrothermal crosslinked gelatin film using a 50 Hz oxygen glow discharge. Surf Coat Technol 205:S133–S138
33. Prasertsung I, Kanokpanont S, Mongkolnavin R, Wong CS, Panpranot J, Damrongsakkul S (2012) Plasma enhancement of in vitro attachment of rat bone-marrow-derived stem cells on cross-linked gelatin films. J Biomater Sci Polym Ed 23:1485–1504
34. Prasertsung I, Kanokpanont S, Mongkolnavin R, Wong CS, Panpranot J, Damrongsakkul S (2013) Comparison of the behavior of fibroblast and bone marrow-derived mesenchymal stem cell on nitrogen plasma-treated gelatin films. Mater Sci Eng C 33:4475–4479
35. Amornsudthiwat Phakdee, Mongkolnavin Rattachat, Kanokpanont Sorada, Panpranot Joongjai, Wong Chiow San, Damrongsakkul Siriporn (2013) Improvement of early cell adhesion on thai silk fibroin surface by low energy plasma. Colloids Surf B 111:579–586
36. Tay WH, Yap SL, Wong CS (2014) The electrical characteristics and modeling of a filamentary dielectric barrier discharge in atmospheric air. Sains Malaysiana 43(4):583–594
37. Tay WH, Kausik SS, Wong CS, Yap SL, Muniandy SV (2014) Statistical modelling of discharge behavior of atmospheric pressure dielectric barrier discharge. Phys Plasmas 21:113502
38. Tay WH, Kausik SS, Yap SL, Wong CS (2014) Role of secondary emission on discharge dynamics in an atmospheric pressure dielectric barrier discharge. Phys Plasmas 21:044502
39. Ramasamy Rajeswari K, Rahman Noorsaadah A, San Wong Chiow (2001) Effect of temperature on the ozonation of textile waste effluent. Color Technol 117:95–97
40. Hashim Siti Aiasah, San Wong Chiow, Abas Mhd Radzi, Hj Khairul Zaman, Dahlan (2007) Feasibility study on the removal of Nitric Oxide (NO) in gas phase using dielectric barrier discharge reactor. Malays J Sci 26:111–116
41. Hashim SA, Wong CS, Abas MR, Hj Dahlan KZ (2010) Discharge based processing systems for nitric oxide (NO) remediation. Sains Malays 39:981–987
42. Bhai Tyata Raju, Prasad Subedi Deepak, Rajendra Shrestha, San Wong Chiow (2013) Generation of uniform atmospheric pressure argon glow plasma by dielectric barrier discharge. Pramana 80:507–517
43. Lee YS, Bora B, Yap SL, Wong CS (2012) Effect of ambient air pressure on synthesis of copper and copper oxide nanoparticles by wire explosion process. Curr Appl Phys 12:199–203

44. Wong CS, Bora B, Lee YS, Yap SL, Bhuyan H, Favre M (2012) Effect of ambient gas species on the formation of cu nanoparticles in wire explosion process. Curr Appl Phys 12:1345–1348
45. Bora B, Wong CS, Bhuyan H, Lee YS, Yap SL, Favre M (2012) Impact of binary gas on nanoparticle formation in wire explosion process: an understanding via arc plasma formation. Mater Lett 81:45–47
46. Bora B, Wong CS, Bhuyan H, Lee YS, Yap SL, Favre M (2013) Understanding the mechanism of nanoparticle formation by wire explosion process. J Quant Spectrosc Radiat Transfer 117:1–6
47. Bora B, Kausik SS, Wong CS, Chin OH, Yap SL, Soto L (2014) Observation of the partial reheating of the metallic vapor during the wire explosion process for nanoparticle synthesis. Appl Phys Lett 104:223108